LIVING IN SPACE

Peter Smolders

LIVING IN SPACE

A Handbook for Space Travellers

Translated by Sidney Woods

TAB *AERO*

Blue Ridge Summit, Pa. 17214

Original Dutch Text © 1985 Romen Luchtvaart, Uniboek, bv.
This English Translation © 1986 Sidney Woods.

First English Language edition published in 1986 by Airlife Publishing.

This U.S.A. edition published in 1986 by Tab/Aero,
Blue Ridge Summit, Pa. 17214.

Library of Congress Cataloging-in-Publication Data

Smolders, P. L. L.
    Living in space.

    Translation of: Wonen in de ruimte.
    1. Space colonies.   I. Title.
TL795.7.S6413   1986   629.45   86-5863
ISBN 0-8306-8480-8 (pbk.)

Printed in Spain.
IHASA, Mallorca, 51. Barcelona

CONTENTS

HAVE A PLEASANT JOURNEY!

Space travel is becoming more and more within the reach of ordinary human beings. People just like you and me . . . This has been a gradual development. In the 1960s highly trained people went into space for the first time — singly or in very small numbers, and only for a few weeks at a time. In the 1970s, almost unnoticed, came the era of living in space; the first small space stations were launched and occupied for periods of up to several months. Then, in the 1980s, a different means of transport became available — the space shuttle — which made it possible to assemble a space station from separately launched blocks, and to maintain a regular service between Earth and such an outpost in space. In the 1990s space will be within the reach of people who have not followed an astronaut's training course, for those who — for personal or commercial reasons — wish to stay in space for a short or even a long period. This development is analogous to what has already taken place in the field of aviation.

So, space travel for all. Hence this handbook, which will familiarize you with many different aspects of the journey to your destination in space, your stay there, and your return to Earth. **Living in Space** is based upon the situation as it will be in 1995. The book is founded upon the American space shuttle as a means of transport and on a 'standard' space station. In 1995 different types of space stations will circle the Earth in a variety of orbits, but in principle the problems and possibilities in one space station will not be essentially different from those in another.

Have you by now 'had it' on good old Mother Earth? Then get ready to leave her, even if only for just a short while. And don't forget to put this book in your luggage.

Have a pleasant journey
Peter Smolders

Acknowledgments

My gratitude goes to the numerous establishments and people who assisted me with collecting the material for this book. At the risk of forgetting somebody, I will attempt to name them — in no particular order:

NASA Headquarters, Washington: Debbie Rahn, Charles Redmond, William O'Donnel, Miles Waggoner; NASA Kennedy Space Centre: Ed Harrison, Mark Hess and Betty McNaughton; NASA Houston: Terry White, Lisa Vazques and Mike Gentry; Rockwell International: Sue Cometa and Dick Barton; Grumman: Patricia Andrews and Ben Kovit; Boeing: Donna Mikov and Don Brannon; Lockheed: Steve Pehanich: ESA: Jaqueline Gomérieux, Heidi Graf, Annemarie Hoyet and Mieke Bruens.

In the sphere of Russian space travel I received much assistance from Valery Kulishov, Vladimar Kramarienko, Tamara Shagnazarova, Vladimir Molchanov, Yuri Goloviatienko, cosmonaut Alexander Serebrov, Professor Oleg Gazenko, Professor Andrei Kapitsa, and many others.

Finally, my thanks to my friend Coen Benraad, who prepared many drawings for this book.

PART I: SPACE SHUTTLE

1 JOURNEY INTO SPACE

In the 1990s a journey to a space station, orbiting the Earth at a height of a few hundreds of miles, will no longer be unusual. But it will remain interesting and exciting, particularly a stay lasting for several weeks or months in' a completely man-made world where everything and everybody is weightless . . .

You will travel to the space station in the space shuttle, the American space vehicle. It starts like a rocket, flies like a spacecraft and lands like a glider. There are three of these space vehicles: **Columbia, Discovery** and **Atlantis.**

You will embark at Cape Canaveral on the east coast of Florida, where the shuttle has been readied on the launching pad. Shortly after a deafening start, which will increase your heart rate considerably, the booster rockets will be jettisoned, and a little later the same will happen to the shuttle's large external fuel tank. Then you will go into orbit around Earth to rendezvous with the space station.

After having lived for a while on board the space station you will return to our planet with her inexorable gravity. The shuttle uses its retro rockets and enters the atmosphere. The friction of the air ensures more braking, and shortly afterwards the space vehicle lands on Cape Canaveral's runway, later to leave with a fresh load and new passengers to travel once again into space. Space travel has become a routine matter, but if you have never been in space it is advisable to acquaint yourself with weightlessness, an experience that provides one surprise after another . . .

> Your journey in a nutshell:
> 1. At the launch site
> 2. Launch
> 3. Boosters jettisoned
> 4. External fuel tank jettisoned
> 5. In orbit around Earth
> 6. Docking with the Space Station
> 7. Return from orbit
> 8. Entering the Earth's atmosphere
> 9. Descent to Florida
> 10. Landing
> 11. Boosters with parachutes land in sea
> 12. Fuel tank drops into the ocean
> 13. Boosters return to the launch site

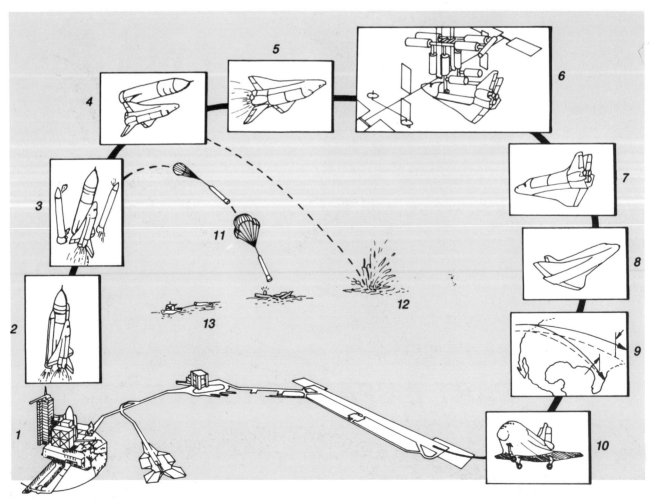

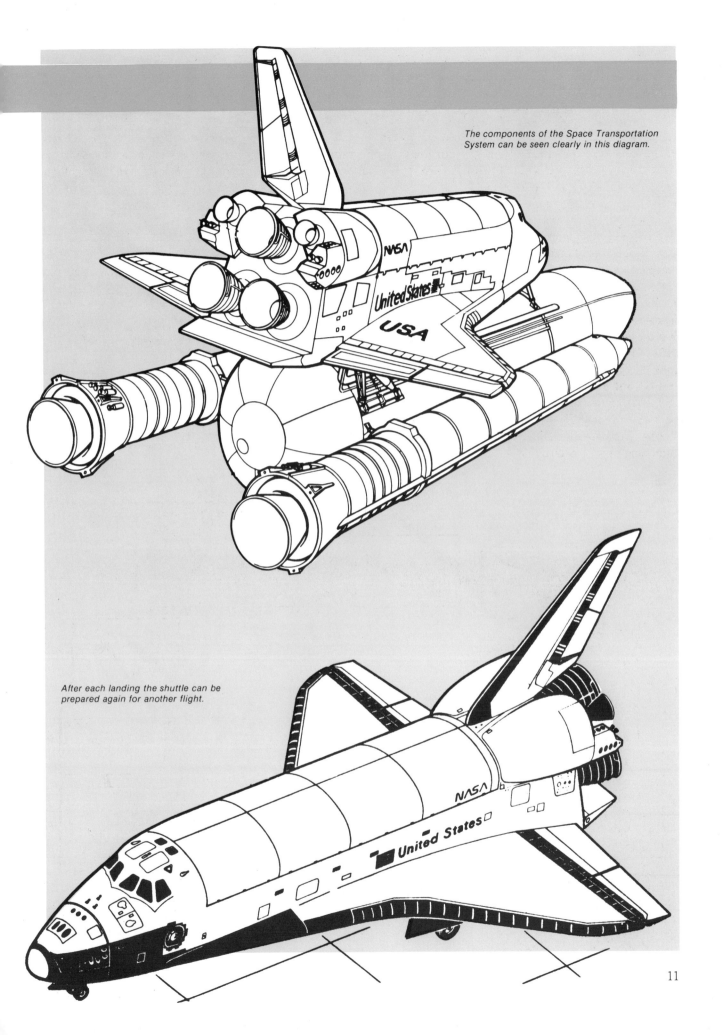

The components of the Space Transportation System can be seen clearly in this diagram.

After each landing the shuttle can be prepared again for another flight.

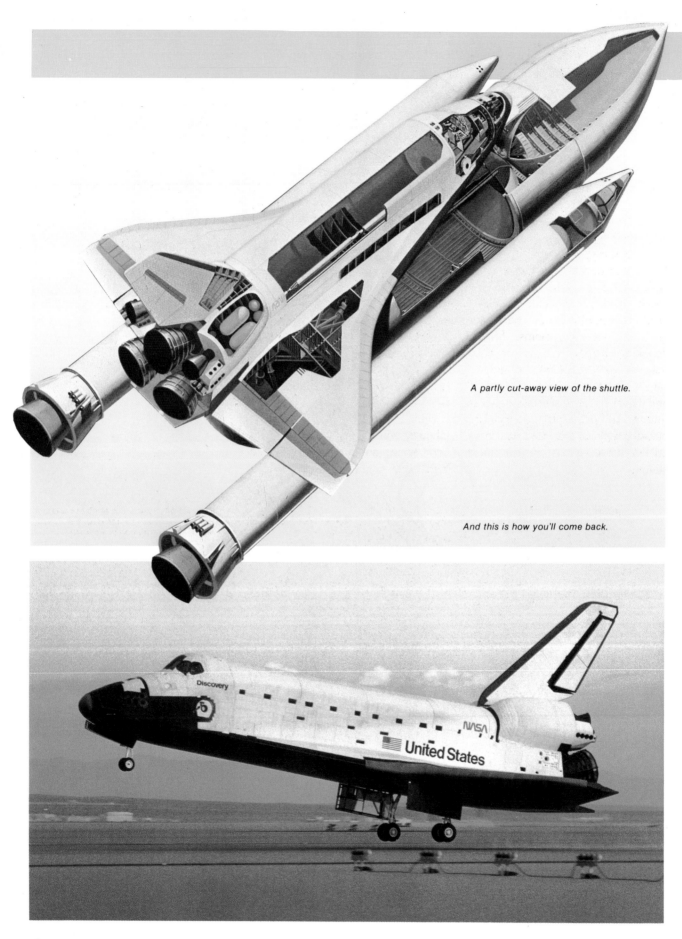

A partly cut-away view of the shuttle.

And this is how you'll come back.

Just about anyone who is bodily fit can make a journey in the space shuttle and visit a space station. The launch and the return trip will not be too traumatic; you will become three times heavier than you normally are, but this is only for a short while and can easily be tolerated because you will be in specially designed seats. The weightlessness — which you will experience during several weeks or perhaps even months on board the space station — takes a little getting used to, but it does not offer unsurmountable problems. Elderly people will be bothered less, because their sense of balance is not as sensitive as that of younger persons, and therefore will not be influenced as much. In principle, everybody between eight and eighty is welcome on board the shuttle and the space station.

Who have so far travelled in the shuttle? Pilots, of course, but also scientists, teachers, members of the American Senate, journalists, painters and draughtsmen, a pop artist, several well-known TV personalities, an Arabian Prince. So why not you? Come on, gather up your courage and put on the flying suit supplied by NASA, put on your safety helmet and go on board. There is no need to put on a space-suit, which is heavy and impedes your movements. Within the shuttle a normal atmosphere prevails with normal pressure. The temperature is pleasant, the seats are comfortable. In fact, there is not much difference between flying in a shuttle or flying in a normal aircraft except that, for the take-off, you lie on your back with your knees bent, your toes a little higher than your head. But before you

reach that phase, there are a few things to think about and to do. Not that these should be difficult.

First, the countdown: you are about to commence the most fascinating journey of your life . . .

This vehicle takes the seven or eight shuttle occupants from the crew centre to the launch pad, and back again upon their return.

During a flight you can wear a headset if you want to speak to people on the ground or with crew members somewhere else in the shuttle or the Space Station. It has a small earplug and a miniature microphone. The combination is connected to a small box which is fastened to your clothing by a clamp. Communication is by radio; you are not kept on a leash.

1. Connecting plug
2. Microphone
3. Adjustable headband
4. Earplug
5. Clamp
6. Volume control
7. Frequency selector switch
8. Protecting cover for volume control
9. Connections for earphone/microphone
10. Antenna lead

A space-suit is not necessary during a journey on the space shuttle – just the standard outfit as made available by NASA and, of course, during launch and landing, a safety helmet.

Your journey starts from the famous Kennedy Space Center in Florida, from where the first people to travel to the Moon also departed. There is also a special shuttle base for military flights, in Vandenberg, California but we'll forget about that one for the time being. Much of what was used in the years of the Apollo missions is still serving in the era of the space shuttle.

Launch site 39 has been modified for use by the shuttle, which is much shorter and more compact than the Saturn 5 Moon rocket. But the huge Vehicle Assembly Building (VAB), for a long time the largest building in the world, still serves, and the same applies to the 'crawler', the biggest freight truck in the world, which used to move the Moon rockets to the launch site, and now transports the space shuttle. New, however, is the special building in which the shuttle is serviced after each trip and is prepared for the next: the Orbiter Processing Facility (OPF). Prior to your journey you will undoubtedly have had an opportunity to look around the Kennedy Space Center thoroughly (it covers more than 140,000 acres). Even if you have no wish to fly from there, it is well worth looking at. There are daily excursions taking you around by bus, and at the most interesting spots you may de-bus to take photographs.

Orbital angles

From KSC the shuttles are launched roughly in an easterly direction, profiting from the rotational direction of the Earth from west to east. The angle formed by the ultimate orbit of the shuttle and the Equator may vary between plus 57 degrees (in a northerly direction) and minus 39 degrees (in a southerly direction). Usually an angle of 28.5 degrees is used, which is the latitude of the Cape. The shuttle can thus lift the maximum possible payload, approximately 29.5 tonnes. Using different angles, the shuttle would not profit so much from the rotational speed of the Earth (because the direction deviates too much), and the useful payload carried would therefore be less. The space station in which you'll be living for several weeks or months also orbits along the track that makes an angle of 28.5 degrees to the Equator: it is the most obvious orbit for launchings from the Kennedy Space Center.

The OPF building

After each flight the spacecraft goes to the OFP building. This is a kind of

The majestic Vehicle Assembly Building in which the shuttle combinations are assembled. The space vehicle is driven from the OPF building to the VAB.

very large hangar, in which two orbiters (spacecraft) can be serviced simultaneously. The OPF has two high servicing bays, each 198ft (60m) long, 150ft (46m) wide, and 95ft (29m) high. It is made up of platforms, girdling and covering the orbiter, combined with a 27-tonne gantry crane. The large bay between the two high bays is 234ft (71m) long, 100ft (30 m) wide and 25ft (7.5m) high. The high bays in which the shuttles are serviced have a system of trunking, which holds the cables for electricity and communications, and also for the hydraulic systems and the gas supplies.

The long bay holds the electronic apparatus, various machines, and the repair installation for the heat-resistant tiles. It also accommodates the necessary offices and control rooms.

The VAB

The Vehicle Assembly Building (VAB) is the nerve-centre of Launch

site 39. This enormous building is 528ft (160m) high, 720ft (218m) long and 520ft (158m) wide. In this building, which with its volume of 3,665,000 cubic metres is one of the largest buildings in the world, two space shuttles can be prepared simultaneously for flight. The shuttle combination is built-up on a mobile launching platform which is transported by the immense crawler to the launch pad. First the two solid fuel rockets are placed in position, next the huge external tank is connected to these boosters, and finally the spacecraft itself is installed against the tank. It goes without saying that for this assembly job tremendous cranes (up to 227 tonnes lifting capacity) are available.

If the VAB could be called the nerve-centre of the Launch site 39, then the Launch Control Center (LCC) must be the brain! The LCC is a four-storey building (very low compared to the VAB) which — looking from the launch pad — lies to

Top, left: After having been joined up with its large external tank and its two booster rockets, the shuttle is about to leave the VAB.

The three large openings in the mobile launch pad are for the unobstructed passage of the exhaust gases.

Opposite page: Standing upon the mobile launch platform, the shuttle leaves the VAB. In the foreground, centre, is the Launch Control Center.

the left of the VAB, linked by a corridor.

Compared with the Apollo era, there are quite a few changes in the LCC. For the launch of an Apollo rocket 450 persons were required, but for a space shuttle start no more than 45 persons are needed in the LCC.

Launch platform

The shuttle, having arrived from the OPF building, will be assembled in the VAB on a mobile launch platform, also originating from the Apollo era but since modified extensively. This platform is two storeys high 25ft, (7.6m), 161ft (48.8m) long and 136 ft (41.1m) wide. It is made of 6 inch (15.2cm) thick steel. When it is parked to the north of the VAB, inside the VAB itself, and on the launch pad, the launch platform rests on six thick columns, 22ft (6.7m) high.

The platform has three large openings for the passage of the exhaust gases produced by the shuttle's engines: two for the solid fuel boosters, and one for the shuttle's three main engines. The openings for the booster exhaust gases measure 42ft (12.8m) by 20ft (6m). The exhaust opening for the shuttle's main engines

(which produce far less thrust than the two boosters) is 34ft by 31ft (10.4 by 9.5m).

The platform also houses a number of rooms holding all kinds of electronic systems, test-apparatus and installations for tanking fuel.

Unloaded, the mobile launch platform (MLP) weighs 3,733 tonnes. The total weight of a platform plus an unfuelled shuttle is 4,989 tonnes. Once the shuttle is fuelled the total all-up weight is no less than 5,761 tonnes.

In each of the openings beneath the solid-fuel rockets, four supports have been installed on which the complete shuttle combination rests. These supports are 5ft (1.5m) high and each has a diameter at its base of 4ft (1.2m).

Conspicuous are the two thick service masts to the left and right of the exhaust opening for the main engines, near the shuttle's tail

assembly. These so-called TSMs (Tail Service Masts) ensure the supply of liquids and gas, and the electrical connections, to the ports for hydrogen and liquid oxygen at the rear of the shuttle. The TSMs protect the ground part of the installation against the blast of the launching. On start-up a charge is detonated, causing a weight of more than 9 tonnes to drop. This causes the ground connections to be withdrawn from the spacecraft, at the same time rotating the mast into a bunker-like capsule. As the mast retracts backwards, a pressure cylinder is activated which presses a cover into position, so completely protecting the whole of the structure against the exhaust gases produced by the shuttle's main engines.

Whenever a shuttle is started, a possibility exists that the engines may

have to be stopped after having functioned for a few seconds (of course, only **before** the two·solid-fuel boosters have been ignited, because these cannot be stopped). In such an event a fire could start on the launch platform, because there will be a considerable quantity of highly inflammable hydrogen still escaping from the exhausts. For this reason the platform has been equipped with 22 water-sprinklers, which are supplied by a 6 inch (15cm) diameter pipe and can spray the underside of the shuttle, on and around the engine exhausts, with 2,000 gallons (9,460 litres) of water per minute.

The largest truck in the world, the 'crawler', carries the mobile launch pad with the shuttle on it.

The 'Crawler-Transporter', abbreviated to 'crawler'. The parts are identified as follows: 1: driver's cabin; 2. mechanical link between crawler and mobile launch pad; 3. one of the eight tracks (each link weighs one tonne).

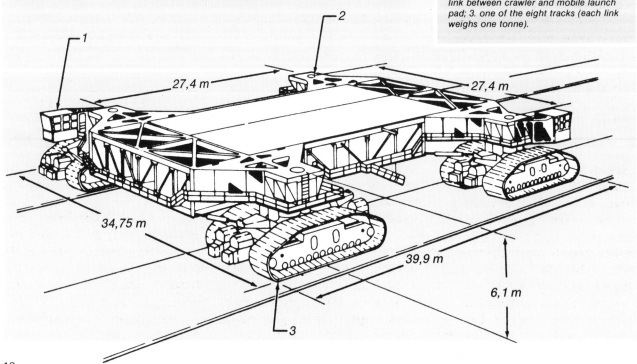

27,4 m   27,4 m

34,75 m

39,9 m

6,1 m

The 'crawler'

The Crawler Transporter (two are available), used for transporting the shuttle standing on the launch platform to the launch site, is 128ft (39m) long and 115ft (34.7m) wide. The crawler moves along on eight enormous tracks (four sets of two pairs), each 10ft (3m) high and 41ft (12.5m) long. When loaded, the crawler attains a speed of no more than 1 mph (1.6 km per hour) and itself weighs 2,721 tonnes. A hydraulic system keeps the upper surface of the crawler exactly horizontal. This is necessary because of the 5 degree gradient leading to the launch site. This system is also used during the positioning of the launch platform in the VAB and on the launch site.

Each crawler is driven by two diesel engines, each of 2,750 kilowatts. These, in turn, drive four 1,000 kilowatt generators which supply electrical power to the 16 electric motors driving the tracks.

The crawler moves with its load along a specially constructed concrete road between the VAB and one of the two launch sites. The distance to Launch site 39A is 3.4 miles (5.5 km), and to Launch site 39B 4.2 miles (6.8km). The road surface has a thickness of 7ft (2.1m). Its top layer is gravel, 8 inches (20cm) thick on the curves and 4 inches (10cm) thick on the straight parts.

Launch sites

Launch sites 39A and B together cover an area of about 165 acres and lie about 50ft (15m) above the surrounding land. The concrete platform of each launch site measures 392ft by 326ft (119 by 99m). Each platform has a fixed launch tower and a rotating service structure (RSS), which is turned away before the shuttle is launched. The RSS can, for

The shuttle moving uphill to the launch site.

instance, be used to load or unload freight into and from the shuttle. The fixed launch tower holds fuel supply lines and also has an arm which reaches close to the shuttle's access hatch. You will enter the spacecraft via this arm.

The launch site is constructed mainly of steel and concrete. In its lowest parts enormous flame deflectors are installed, for deflecting sideways the exhaust gases from the shuttle's main engines and the two solid fuel boosters. The flame deflectors are of steel, and covered with a heat-resistant ceramic material 5 inches (12.7cm) thick.

At the moment the shuttle lifts off, a tremendous amount of water descends on the launch platform.

This is done to suppress the vibration caused by sound waves bouncing back from the mobile launch platform, because these vibrations can have a destructive effect upon the shuttle and its load. The sound-wave suppression system consists of an enormous tank holding more than 252,000 gallons (1,135,500 litres) and mounted on a tower. The tank is 291 ft (88.4m) high. Just before the shuttle's engine is ignited, the water is released to run through 7ft (2.1m) diameter pipes to the launch platform. Nine seconds after lift-off the flow of

water is at its peak: more than 660,000 gallons (3,000,000 litres) per minute. The sound wave is at its greatest when the shuttle is about 300ft (90m) above the ground. Lower than that, the sound mainly passes away via the flame deflectors; higher than that, the launch installation itself ensures adequate scattering. At about five seconds after lift-off the sound wave is at its greatest; after about 10 seconds the problem is over.

The shuttle's main engines consume liquid fuel — liquid hydrogen, which is burned with the aid of liquid oxygen. The liquid oxygen is stored in a 750,000 gallon (3,406,500 litres) tank in the north-west corner of the launch area. This enormous spherical tank is actually a large thermos flask in which the liquid oxygen is stored at a temperature of less than 90 Kelvin (minus 183

The fire-engine in front of the pad looks insignificant in relation to the crawler creeping upwards on the hill. The launch platform is kept absolutely horizontal during this approach.

Left: Natural and man-made beauty of Launch site 39A.

Top, right: Tiny people and the enormous testimony to their technical capabilities.

Below, right: The mobile launch platform is almost in position. Note the mobile service-gantry on the left. Once the shuttle is in position this gantry moves to embrace the space vehicle completely.

degrees Centigrade). Liquid oxygen (called 'lox' by the technicians) is rather heavy: one litre weighs 2.6lb (1.18kg). The lox is transported to the shuttle's external tank by two pumps with a capacity of 8,327 gallons (37.850 litres) per minute.

The liquid hydrogen, the actual fuel for the shuttle, is stored in a similar spherical tank of 704,000 gallons (3,200,000 litres) capacity positioned in the north-east corner of the launch area. Liquid hydrogen must be stored at a much lower temperature than lox, because its boiling point is lower.

Liquid hydrogen starts to evaporate at temperatures above 20 Kelvin (minus 253 degrees Centigrade) and is very light, only weighing 60 grammes per litre. Therefore, no pumps are necessary to make it flow to the launch site. Evaporators ensure that a small

proportion of the hydrogen is changed into gas, and it is this gas, in the uppermost part of the tank, which builds up sufficient pressure on the liquid, super-cooled hydrogen to make it flow once the valves have been opened. Of course, the pipes through which the liquid hydrogen and oxygen flow have been adequately insulated. They carry the two liquids to the tail end of the spacecraft.

The Space Transportation System (STS) in a vertical position. This system comprises three major components: the spacecraft, the external tank, and two rockets operating on solid fuel. Each of these rockets deliver 11,600 kilo-Newton thrust on take-off, that is, 1,160 tonnes. This is far more than that delivered by the space shuttle's main engines, which each deliver 2,100 kilo-

Before the flight: The Countdown

Time	Event
T-43 hrs	Call to stations. A summons to all control centres and tracking stations
T-38 hrs	Start preparations for fuelling external tank
T-34 hrs	Start evacuation of spacecraft and surrounding installations for the launch (everybody out who has no urgent business there)
T-26 hrs	Everybody away from the launch platform, ready for filling the fuel tanks
T-24 hrs	Loading liquid oxygen into fuel tanks
T-21 hrs 30 min	Loading liquid hydrogen into fuel tanks
T-19 hrs 30 mins	Switch on all navigational apparatus in the spacecraft, and start warming-up period
T-19 hrs	Communications check with tracking stations at Merrit Island and Mission Control, Houston
T-18 hrs	Switch on communications system of the spacecraft
T-15 hrs	Ground crew checks the positions of the switches in the shuttle and prepares service tower for withdrawal
T-12 hrs	Preparation of ventilation cover for gaseous oxygen on external tank and removal of flight-deck platforms.
T-11 hrs	Begin 'Built-in hold'. Countdown clock stops for previously planned period. This is to cater for possible problems. The duration of the hold depends on the time of launching
T-11 hrs	Withdrawal of rotating service tower
T-8 hrs 40 mins	Switch on fuel cells
T-8 hrs	Communication Mission Control: prepare for launching
T-7 hrs 15 mins	Shuttle cargo bay: pump full of nitrogen prior to fuelling
T-6 hrs	Start filling external tank (the large external tank of the shuttle) with liquid oxygen and liquid hydrogen.

Time	Event
T-5 hrs	Adjust navigational instruments and check for security of launch complex and surrounds
T-3 hrs	A two-hour hold begins. External tank is full. Awaken the crew (4 hours 20 minutes prior to lift-off)
T-2 hrs 30 mins	Crew departs in special van for the launch platform
T-1 hrs 55 mins	Crew enters spacecraft
T-20 mins	Shuttle computers ready for launch
T-9 mins	A 10-minute hold begins. Check for and receive 'Go' from Launch Director
T-9 mins	Start automatic countdown (Ground Launch Sequencer)
T-7 mins 30 secs	Withdrawal of access arm from spacecraft
T-5 mins	Start auxiliary power units on the shuttle: safety devices, external tank and booster rockets switched on.
T-3 mins 30 secs	Spacecraft switches to own energy source
T-2 mins 55 secs	Liquid oxygen tank fully pressurized, withdrawal of ventilation cap
T-2 mins 57 secs	Liquid hydrogen tank pressurized
T-31 secs	'Go' from automatic sequencers to shuttle computers to take over automatic countdown
T-28 secs	Start hydraulic power sources (among other things, these service the main engines of the shuttle)
T-10 secs	'Go' for starting main engines
T-6.6 secs	Start shuttle's main engines
T-3 secs	Main engines at 90 per cent thrust
T-0	Clamps off, ignition of solid fuel boosters, and lift-off.

A view of Launch site 39A. On the right is the water tower from which the water is drawn to dampen the noise during take-off.

Extreme left: A unique photograph of 'your' space shuttle on Launch pad 39A.

Left: Right at the top is the ventilation cover which prevents ice-formation by escaping liquid oxygen on the uppermost part of the external tank.

The liquid hydrogen flows through thick pipelines to the external tank of the shuttle. On the right is the water tower.

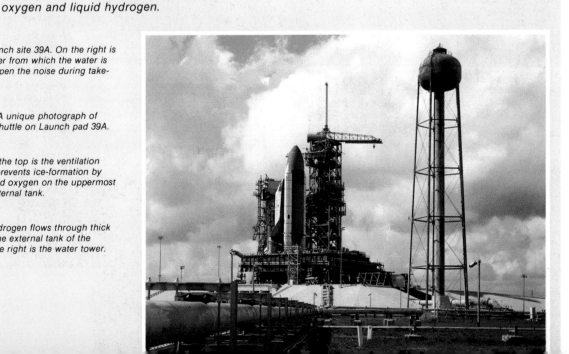

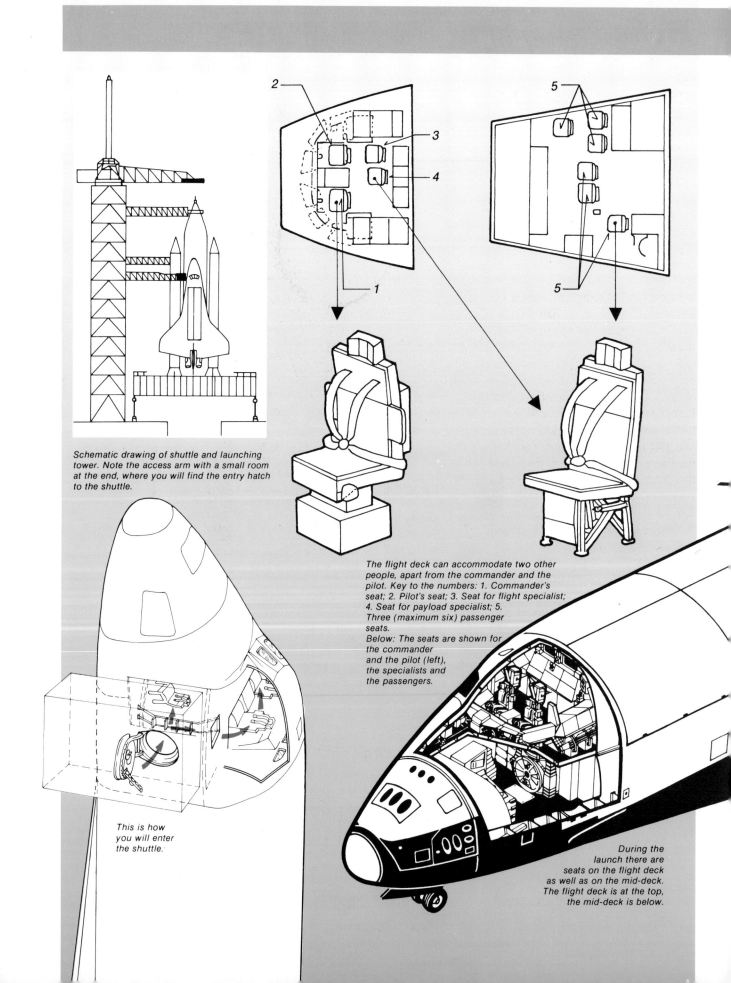

Schematic drawing of shuttle and launching tower. Note the access arm with a small room at the end, where you will find the entry hatch to the shuttle.

The flight deck can accommodate two other people, apart from the commander and the pilot. Key to the numbers: 1. Commander's seat; 2. Pilot's seat; 3. Seat for flight specialist; 4. Seat for payload specialist; 5. Three (maximum six) passenger seats.
Below: The seats are shown for the commander and the pilot (left), the specialists and the passengers.

This is how you will enter the shuttle.

During the launch there are seats on the flight deck as well as on the mid-deck. The flight deck is at the top, the mid-deck is below.

THE SPACE SHUTTLE SYSTEM

The Space Transportation System (STS) in a vertical position. This system comprises three major components: the spacecraft, the external tank, and two rockets operating on solid fuel. Each of these rockets deliver 11,600 kilo-Newton thrust on take-off, that is, 1,160 tonnes. This is far more than that delivered by the space shuttle's main engines, which each deliver 2,100 kilo-Newton (210 tonnes). The spacecraft itself is 122 ft (37m) long and has a wingspan of 79ft (24m). Its weight — without fuel — is 163,170lb (74,000 kg). These figures are comparable to those of a DC-9 aircraft. The spacecraft can, depending on the orbit selected, take a maximum load of 30 tonnes up into orbit around Earth. The cargo bay, closed off by two large doors, is 60ft (18.3m) long and 15ft (4.6m) in diameter. Fuel for the three main engines is supplied by the 155ft (47m) long external tank (diameter 28.7ft (8.7m). At start-up this tank holds 703 tonnes of liquid hydrogen and liquid oxygen. These are in separate compartments: the liquid oxygen at the top, the liquid hydrogen in the lower part. The large tank is the only part of the Space Transportation System which is not reusable. The total combination is 177ft (53.6m) high.

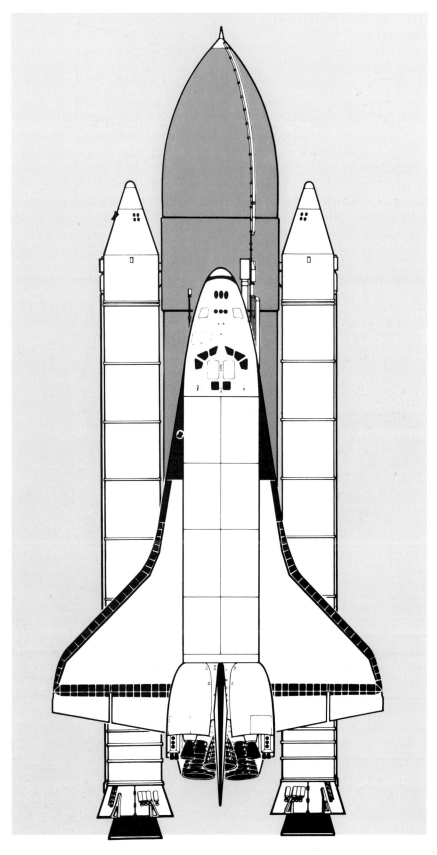

HOT, HOTTER, HOTTEST

The most important technical innovation on the spacecraft is the so-called 'heat-shield'. This consists of 30,900 thick ceramic tiles, which are glued to the spacecraft. At first sight the material looks just like 'tempex', the well-known packing and insulating material. But where tempex disappears like snow in the sun when it gets very hot, this particular material can withstand heat particularly well. The basic ingredients are sand and silicon oxide, which are formed into silica fibres (a glass-like thread). The construction is porous: a tile consists almost wholly of cavities. As a result the tile material has extremely good insulation properties. Heat, once absorbed, can be dissipated quickly: if, for instance, you were to heat a tile with a blowlamp, you could pick it up with your bare hands directly afterwards.

This basic material, after having been cut to a predetermined size (the shape of each tile is calculated according to its position on the shuttle!), is covered with a ceramic layer and then baked in an oven. Heat-resistant tiles with a white top-layer are used on the top surface of the shuttle, where the temperatures during the return to the atmosphere do not rise above 700 degrees Centigrade. Material with a black surface (carbon) is used on the under side of the spacecraft. These tiles can easily withstand a temperature of 1,300 degrees Centigrade. There are 24,100 black and 6,800 white tiles.

Because the ceramic material is rigid, it cannot be glued directly to the aluminium skin of the spacecraft, which contracts and expands with temperature. This is why a material similar to felt is first pasted onto the skin, after which the tiles are attached. The felt absorbs the differences caused by contraction and expansion.

Technicians attaching heat-resistant tiles to the bottom of the shuttle.

The parts of the spacecraft which become hottest during the return are, of course, the nose-cone and the wing leading edges. These rise above 1,300 degrees Centigrade and are therefore covered by a highly efficient heat-resistant and reinforced pure carbon protection system.

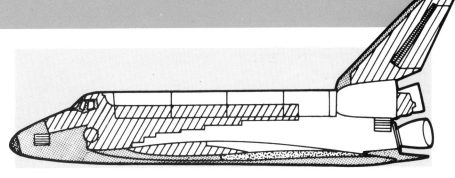

Key to the abbreviations used for ceramic materials:

RRC   Reinforced Carbon-Carbon

HRSI   High-temperature Reusable Surface Insulation

LRSI   Low-temperature Reusable Surface Insulation

FRSI   Flexible Reusable Surface Insulation

REINFORCED CARBON-CARBON (RCC)

HIGH-TEMPERATURE REUSABLE SURFACE INSULATION TILES (HRSI)

LOW-TEMPERATURE REUSABLE SURFACE INSULATION TILES (LRSI)

FLEXIBLE REUSABLE SURFACE INSULATION (FRSI)

METAL OR GLASS

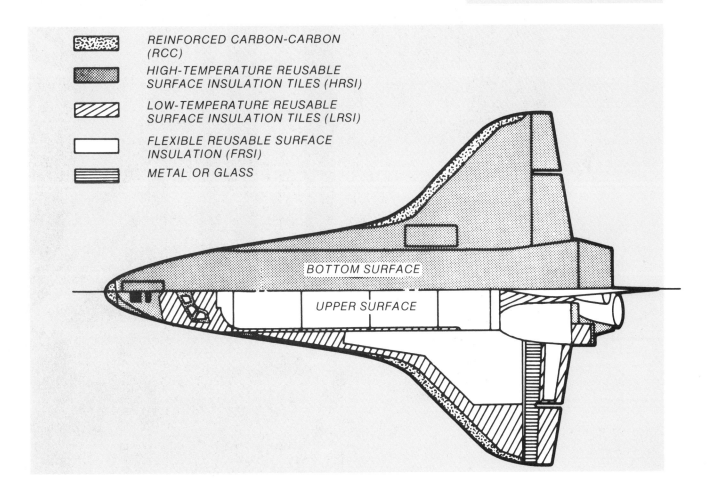

BOTTOM SURFACE

UPPER SURFACE

The space shuttle system, side view:

TOTAL WEIGHT:
Empty: 35,425 kgs
Full: 756,441 kgs

FUEL:
Liquid oxygen: 616,493 kgs
Liquid hydrogen: 102,618 kgs
Total weight: 719,111 kgs

FUEL VOLUME:
Liquid oxygen: 119,214 gallons
(541,482 litres)
Liquid hydrogen: 318,979 gallons
(1,449,905 litres)
Total volume: 438,105 gallons
(1,991,387 litres)

DIMENSIONS:
Liquid oxygen tank: 54ft (16.3m)
Liquid hydrogen tank: 98ft (29.6m)
Intertank (between tanks): 23ft
(6.9m)

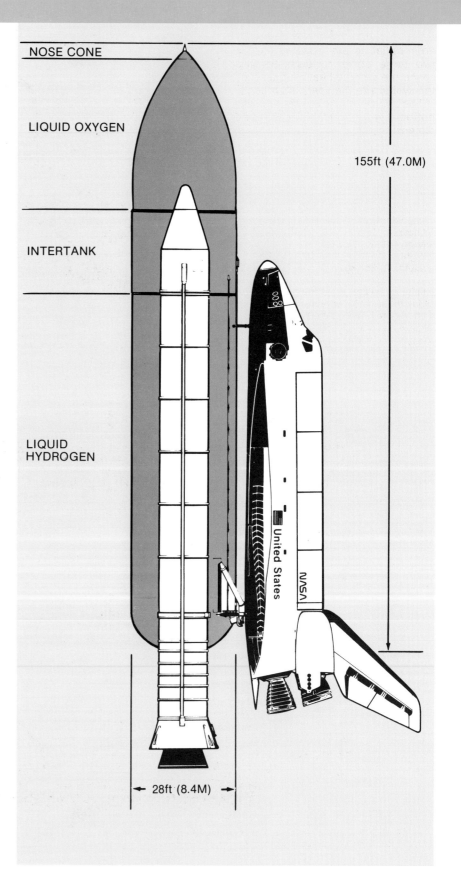

NOSE CONE

LIQUID OXYGEN

INTERTANK

LIQUID
HYDROGEN

United States

NASA

155ft (47.0M)

28ft (8.4M)

Four radiators are installed on the two large doors to dissipate into space the surplus heat produced by the systems in the space shuttle.

The spacecraft front, rear, top and bottom, and the major parts and dimensions.

Dimensions and Weights:
Wingspan: 78ft 5ins (23.79m)
Length: 122ft 10ins (37.24m)
Height: 56ft 11ins (17.25m)
Weight: 74,844 kgs

Minimum heights from the ground:
Body flap (rear): 12ft 2ins (3.68m)
Landing gear door: 2ft 10ins (0.87m)
Nosewheel door: 3ft 5ins (0.90m)
Wingtip: 12ft (3.63m)

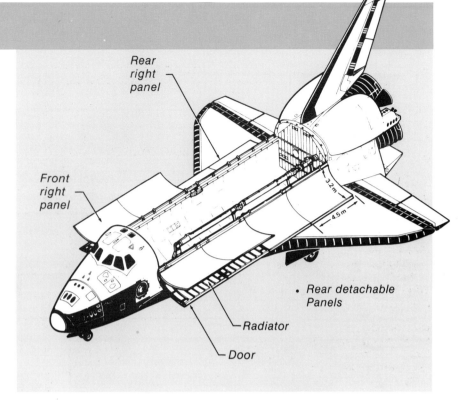

Rear right panel

Front right panel

3.2m

4.5m

• Rear detachable Panels

Radiator

Door

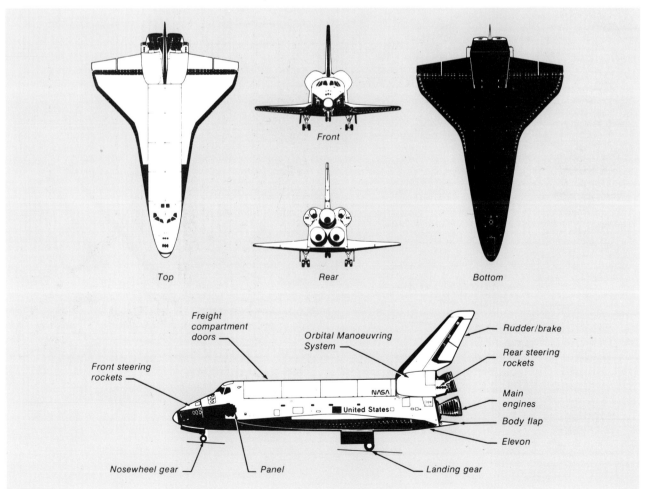

Top

Front

Rear

Bottom

Freight compartment doors

Orbital Manoeuvring System

Rudder/brake

Rear steering rockets

Front steering rockets

NASA

United States

Main engines

Body flap

Elevon

Nosewheel gear

Panel

Landing gear

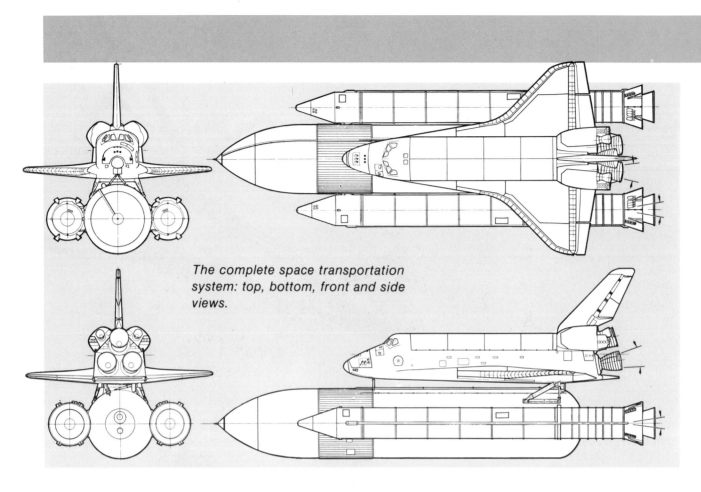

The complete space transportation system: top, bottom, front and side views.

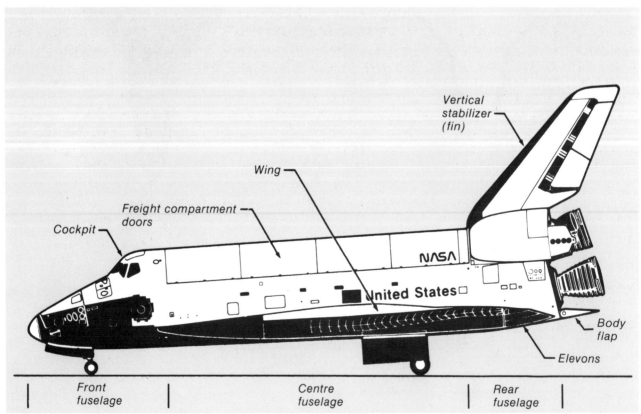

Vertical stabilizer (fin)

Wing

Freight compartment doors

Cockpit

NASA

United States

Body flap

Elevons

Front fuselage

Centre fuselage

Rear fuselage

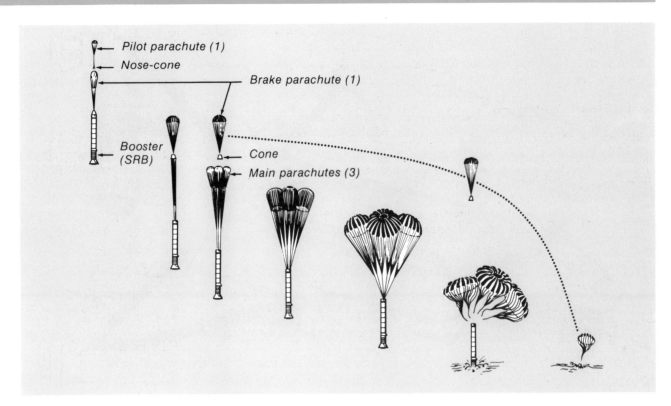

Pilot parachute (1)
Nose-cone
Brake parachute (1)
Booster (SRB)
Cone
Main parachutes (3)

The Boosters

The solid fuel rockets (usually called boosters) which, together with its main engines, will take your spacecraft up, are jettisoned at a height of 30 miles (50 km), by which time they are empty.

Of course, after having been released they continue to fly for quite some distance, but after about 4 minutes they drop back to Earth, during which they achieve a maximum speed of 2,888 mph (4,650 km/hr). At an altitude of 2.9 miles (4.69 km) the nose-cone is jettisoned (the height being determined by a barometric pressure switch) after which a pilot parachute is deployed.

This 'chute in turn pulls out the brake parachute, which has a diameter of 54ft (16.5m). This opens gradually and retards the boosters to such an extent that the three main parachutes can be opened to take over at an altitude of 1.2 miles (2 km); each has a diameter of 115ft (35m) and the lines to the booster are 172ft (52m) long. The boosters ultimately plunge into the ocean at a speed of 60 mph (95 km/hr).

They are pumped full of air by a ship and then towed back to Cape Canaveral, to be prepared anew for yet another flight by the space shuttle.

Drawing left: the space shuttle, side view:		Centre fuselage	length 60ft 5ins (18.3m) width 17ft 2ins (5.2m) height 13ft 2ins (4.0m)
Total length	122ft 10 ins (37.24m)		
Height	56ft 11ins (17.25m)	Front fuselage	71.5 cu.m
Vertical stabilizer fin	26ft 5ins (8.01m)	Cargo bay doors	length 60ft 5ins (18.3m) diameter 15ft 2ins (4.6m) surface 487.2sq.ft (148.6 sq.m)
Body flap	length 135.5 sq.ft (12.6 sq.m) width 20ft 2ins (6.1m)	Wing	length 60ft 5ins (18.3m) max. thickness 4ft 11ins (1.5m)
Rear fuselage	length 18ft 2ins (5.5m) width 22ft 1in (6.7m) height 20ft 2ins (6.1m)	Ailerons	13ft 10ins (4.2m) 12ft 6ins (3.8m)

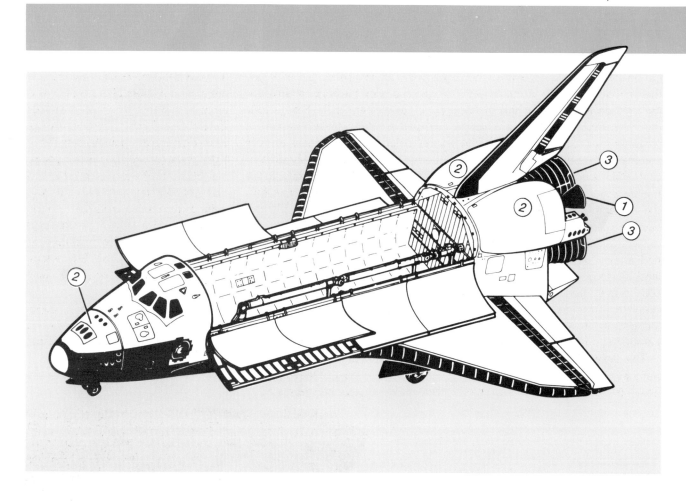

Propulsion

The space shuttle's propulsion system consists of the three main engines, the solid fuel rockets, the external tank, the Orbital Manoeuvring System (OMS) and the Reaction Control System (RCS). The OMS takes the spacecraft on to its final orbit, permitting changes from one orbit to another, takes care of the rendezvous with another spacecraft or with a space station, and the return from orbit (braking manoeuvre). The RCS controls the spacecraft's attitude (rolling, yawing and pitching) while the spacecraft is at an altitude of more than 13 miles (21 km) above Earth.

ORBITAL MANOEUVRING SYSTEM
Two engines, each with a thrust of 26.688 newtons.
Fuel: Monomethyl Hydrazine and Nitrogen Tetroxide.

REACTION CONTROL SYSTEM
One system in the nose, two in the rear.
38 primary thrusters (14 in front, 12 in the rear), each with a thrust of 3870 newtons.
6 vernier engines (two in the front, four in the rear) of 111 newtons each.
Fuel: Monomethyl Hydrazine and Nitrogen Tetroxide.

MAIN ENGINES
Three of them, producing a thrust of 2.100.000 newtons each.
Fuel: Liquid Hydrogen and Liquid Oxygen.

The OMS is indicated in the illustration with 1, the RCS 2 and the main engines with 3.

Right: The flight deck (top) and the mid-deck can be seen here in this cut-away drawing. The two seats for the commander and the pilot are not occupied because the shuttle is now in orbit around Earth.

INSIDE THE SHUTTLE

On board the shuttle you will travel in the 'mid-deck', which is underneath the flight deck. On the flight deck are the commander, the pilot and a maximum of two specialists. The commander and the pilot fly the shuttle (if it is not flown by computer), the mission specialist knows all about the systems in the shuttle and the payload specialist is the expert on the cargo on board. The crew will have had an intensive training lasting at least three years, including at least 600 hours' practice on a flight simulator at Houston, which is identical to the shuttle. For you as a passenger, a comforting thought.

The flight deck has an incredible number of instruments and switches. This is, among other reasons,

because all important systems are triplicated. For instance, should the electrical system fail during your journey to the space station, a second, identical, system takes over at once. And in case this should also fail (a rather remote possibility, as we are talking of complete and fully independent systems) then there is yet a third system available.

The mid-deck, where you have been allotted a seat, is 13ft (4m) long; at the rear it is 12ft (3.7m) wide and at the front (towards the nose) 9ft (2.7m). But much of this space is occupied by apparatus and storage lockers, so you will not be able to see much of the actual walls. To the left of your seat is the toilet, of the same type that you will find in the space station.

To your right and in front of you there are a great number of drawers, which among other things contain the meals for flights of longer duration. At the rear you'll find the stairs leading to the flight deck (never used when you're weightless, at the most you'll need to pull yourself up a little), and also behind you is the access panel to the airlock (positioned in the cargo bay) through which you must pass on your way to the space station. The airlock is also used for possible space-walks outside the shuttle.

These and the following pages contain additional information about a number of systems on board the shuttle.

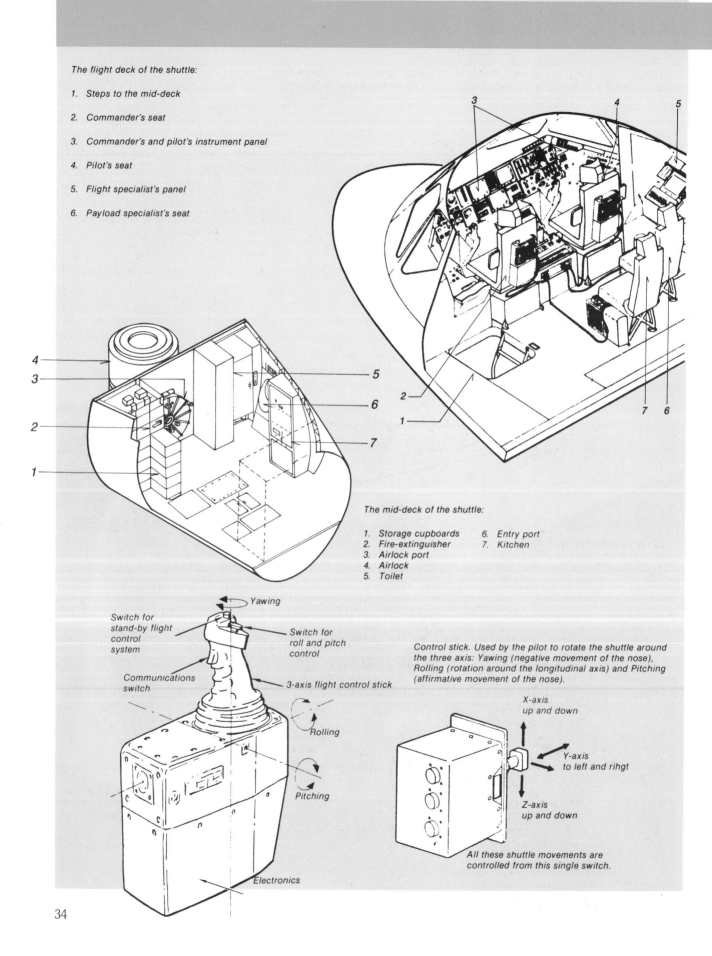

The flight deck of the shuttle:

1. Steps to the mid-deck

2. Commander's seat

3. Commander's and pilot's instrument panel

4. Pilot's seat

5. Flight specialist's panel

6. Payload specialist's seat

The mid-deck of the shuttle:

1. Storage cupboards
2. Fire-extinguisher
3. Airlock port
4. Airlock
5. Toilet
6. Entry port
7. Kitchen

Yawing

Switch for
stand-by flight
control
system

Switch for
roll and pitch
control

Communications
switch

3-axis flight control stick

Rolling

Pitching

Electronics

Control stick. Used by the pilot to rotate the shuttle around
the three axis: Yawing (negative movement of the nose),
Rolling (rotation around the longitudinal axis) and Pitching
(affirmative movement of the nose).

X-axis
up and down

Y-axis
to left and rihgt

Z-axis
up and down

All these shuttle movements are
controlled from this single switch.

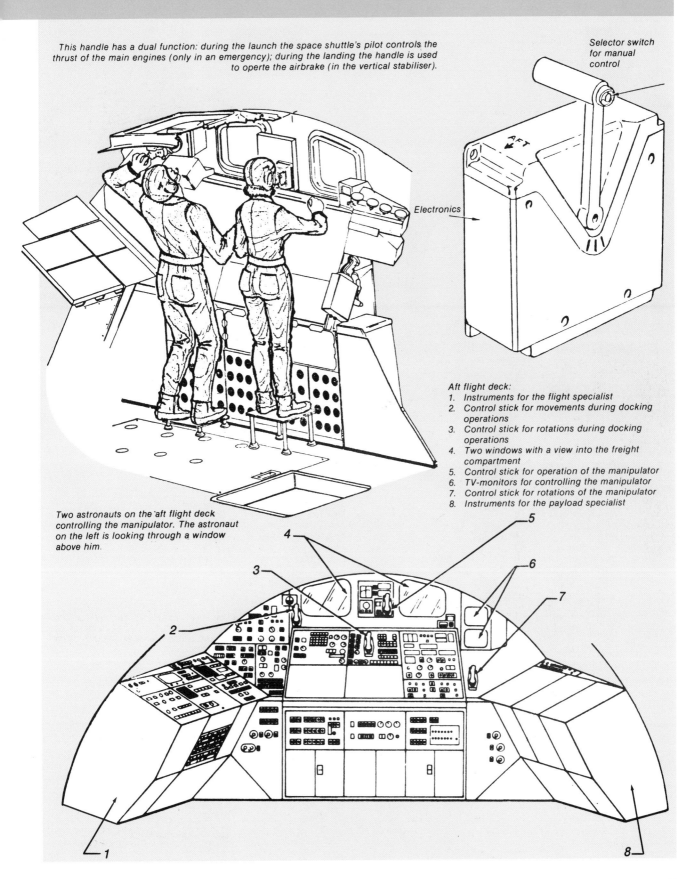

This handle has a dual function: during the launch the space shuttle's pilot controls the thrust of the main engines (only in an emergency); during the landing the handle is used to operte the airbrake (in the vertical stabiliser).

Selector switch for manual control

Electronics

Aft flight deck:
1. Instruments for the flight specialist
2. Control stick for movements during docking operations
3. Control stick for rotations during docking operations
4. Two windows with a view into the freight compartment
5. Control stick for operation of the manipulator
6. TV-monitors for controlling the manipulator
7. Control stick for rotations of the manipulator
8. Instruments for the payload specialist

Two astronauts on the aft flight deck controlling the manipulator. The astronaut on the left is looking through a window above him.

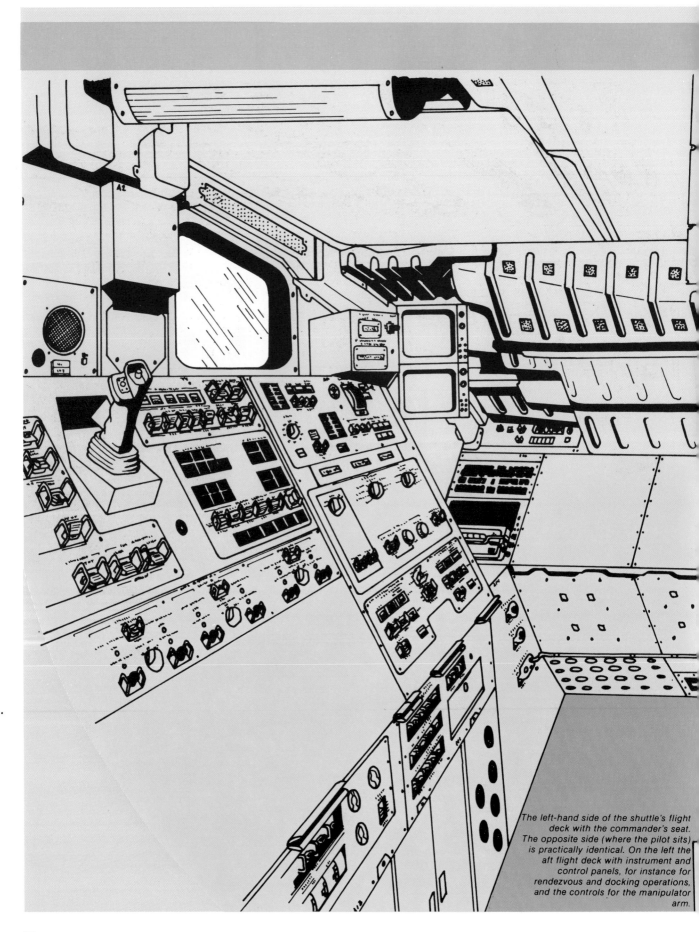

The left-hand side of the shuttle's flight deck with the commander's seat. The opposite side (where the pilot sits) is practically identical. On the left the aft flight deck with instrument and control panels, for instance for rendezvous and docking operations, and the controls for the manipulator arm.

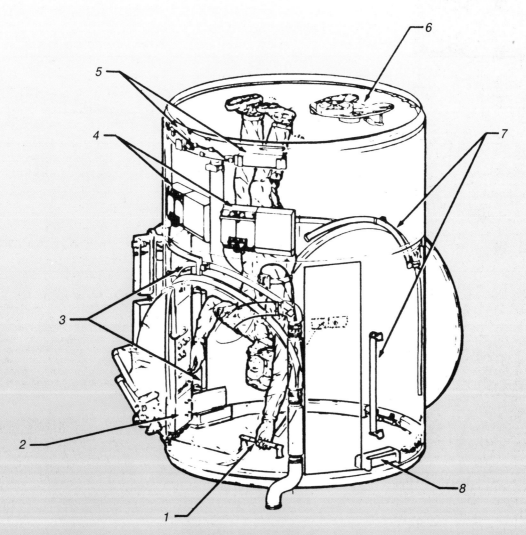

The space shuttle's air-lock

1. Handgrip

2. Control panel

3. Handgrips

4. Portable oxygen system

5. Lights

6. Footholds

7. Handgrips

8. Lights

THE OUTWARD JOURNEY

5-4-3-2-1-ignition!

The start-up. You will be lying on your back on one of the seats in the shuttle, on the mid-deck, and via your headset will have followed the countdown. Fifty metres below, you hear a rumbling noise like a fast-approaching thunderstorm. You know that this is the moment when all hell is breaking loose down there, as one hundred million horsepower combine to lift two thousand tonnes of aluminium, titanium, electrical circuits, computers, fuel and human beings. Amidst a cauldron of whirling clouds of steam and cascades of fire your journey has commenced — dazzling and deafening — but in the beginning you notice only a little of all this. Then, slowly, as if reluctant to begin its long climb into space, the shuttle increases its speed, and the effect of the acceleration grows. However, this will not be too unpleasant: no more than 3g — three times your weight on Earth. During the Apollo flights the astronauts sometimes felt more than eight times heavier than normal, but the shuttle appears to be a gentle giant.

About one minute after lift-off the pressure on your body becomes greatest. Undoubtedly slightly nervous, you probably won't realize that you are, at that time, already 6.8 miles (11 km) above the swamps of Florida. The aircraft's domain is now below you. Barely one minute later, while the pressure is decreasing, you hear in your headset that the boosters, the big rockets, have completed their task and have been jettisoned. This implies an altitude of 29 miles (47 km) and a speed of almost 3,100 mph (5,000 km/hr).

Below: This is how your space flight starts from the Kennedy Space Center.
Left: The two boosters (auxiliary rockets using solid fuel) supply most of the thrust. The exhaust flames of the three main engines of the space shuttle can hardly be seen here. In fact, only water (as steam) issues from these engines, because hydrogen is burned together with oxygen.

The climb is measured in minutes and seconds, but the almost unbearable strain makes it seem like hours. Unbelievable! The cabin clock indicates only nine minutes' flight time when you receive the message that the large, now empty, external fuel tank has been jettisoned.

You are now 75 miles (120 km) above Earth. The space giant leans more on to its back and, propelled by its two OMS (Orbital Manoeuvring System) engines, flashes into its orbit. If you look outside now you'll see the blue Earth suspended in the black heavens. In a great arc your spacecraft has flipped over backwards into space. But up and down have no real significance here: you are now weightless in your safety-straps. Several hours later, after a number of manoeuvres, you'll see through the window a spider-like contraption reflecting flashes of sunlight in your direction: Space Station, the shuttle's destination. The distance to the space station decreases. You can see the crew's quarters, large panels with solar cells supplying energy, a hangar, some space cranes, numerous landing stages. One of the docking hatches approaches you slowly and steadily: the final phase of the rendezvous has commenced. A light bump and the docking is accomplished: you can now begin to transfer. Slowly you float through the open hatch of the airlock for your first meeting with the space station, your new residence in space.

Every launch of the shuttle is more than just an impressive occasion.

On the opposite page: the thin exhaust gases from the three shuttle engines (left) as compared to the thick volumes of fire emitted from the solid fuel rockets.

During the Flight

Time	Event
00.00.00	Launch
00.00.06	Above launch-tower, height 106 metres, speed 120 Km/hour
00.00.07	Shuttle commences to lean over backwards, height 137 metres, speed 123 Km/hour
00.00.11	120 degrees roll, 'heads down'
00.00.30	Roll-manoeuvre completed
00.00.44	Main engines from 100 to 60 per cent thrust
00.01.06	Main engines back to 100 per cent thrust again after achieving maximal acceleration
00.02.00	Boosters burned out
00.02.07	Jettisoning of boosters at 47 Km. height; speed 4.625 Km/hour
00.04.20	Emergency return to Earth no longer possible: 'Negative Return'
00.07.00	Even if two main engines malfunction at this point, an orbit can still be reached
00.07.40	Main engines' thrust reduced to keep excessive load below 3 G's
00.08.28	Main engines' thrust reduced to 65 per cent
00.08.32	Main engines switched off; height 118.5 Km, speed 26.715 Km/hour
00.08.51	Jettisoning of external tank at 118.7 Km height, speed 26.710 Km/hour
00.10.39	OMS-engines ignite (duration 1 minute 10 seconds), orbit of 83 x 250 Kilometers reached
00.49.39	OMS-engines ignited for second correction of orbit (duration 1 minute 31 seconds). Orbit reached now is 250 x 250 Kilometres.

Subsequently some more orbital corrections are carried out (depending upon the type of mission) and rendezvous with the Space Station.

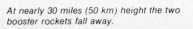

The shuttle leans over backwards, at the same time performing a half-roll.

In the spacecraft, far above the clouds, you leave the domain of normal aircraft far behind you.

At nearly 30 miles (50 km) height the two booster rockets fall away.

Just prior to the shuttle reaching its orbit, the large fuel tank is jettisoned. The tank, which by now is almost empty, returns to Earth and will mostly burn up upon reaching the atmosphere. The remainder drops into the sea.

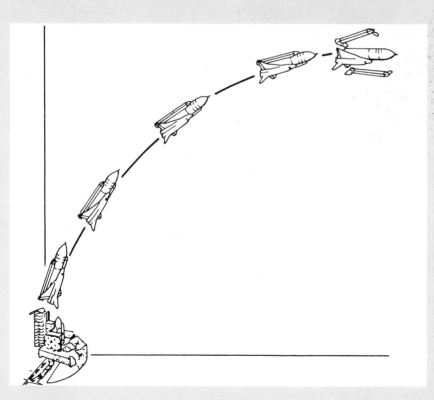

While performing a roll, the shuttle flies into space backwards, as it were, and upside down. This enables the pilots to observe the ground. This flight pattern is also the best for performing a possible emergency return to the launching field.

This is how you will travel in the space shuttle in an orbit around Earth on your way to the Space Station. The two doors along the freight compartment must be open: along the insides are radiators to dissipate the surplus heat into space.

BACK TO EARTH

After some weeks or months, an end has to come to your weightless existence on board the space station — though with regret. The homeward journey is to begin. And for this you must dress specially. You will don a pair of inflatable pants, which press on the lower part of your body and your legs. This stops too much blood flowing away from your head to your legs. During the return journey you will experience a maximum of 1.5g, but the return from zero gravity to one and a half times your normal weight is, of course, quite something. And, what's more, you'll undergo that 1.5g head first; you'll not be lying on your back, as during the launch, Lying on your back, you can more easily bear the extra pressure, because it is spread over a larger surface area.

Even after you get back to Earth it will not be at all easy at first. It will take a few days to get used to.

Two hours before touch-down you will enter the space shuttle again via the airlock on the space shuttle. You'll take the seat allotted to you and wait for things to happen. Of course, you will not forget to fasten your safety-belt and to put on your safety helmet. Through this helmet you can listen in on the communications between the crew of the space station, the shuttle and flight control on Earth.

Twenty minutes after you board, the shuttle gently releases itself from the space station. Looking out of a window (if you're lucky enough to be seated near one) you can see how the Space Station appears to drift away slowly. When the shuttle is several kilometres away from the space station, the crew begins to run through the check-list for the firing of the OMS engines, which in this case act as braking rockets. The run through the check-list takes half an hour.

One hour and fifteen minutes before the landing the shuttle's commander rotates the shuttle so that the two OMS engines point towards the direction of travel. The shuttle is now also upside down, so that the crew can see the Earth. Precisely one hour before the landing the two engines are started and fired for just over two minutes. This

The return journey begins: two small engines, which together form the OMS (Orbital Manoeuvring System), reduce the speed of the shuttle (moving from right to left) in such a way that the shuttle begins to drop in a wide arc back to Earth.

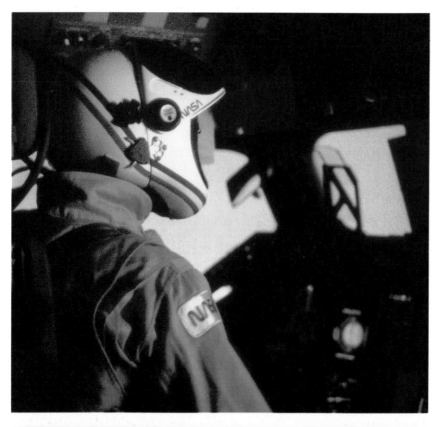

reduces the shuttle's speed by not more than 200 mph (320 km/hr), so that the speed is still 16,684 mph (26,867 km/hr). But this slight reduction in speed is exactly right to take the shuttle out of its orbit: it slowly but surely begins to fall back to the atmosphere, which will take care of the rest of the braking process. The commander now continues to rotate the shuttle until the nose points forward and the craft is moving parallel to the Earth. He then pulls the nose up until an angle of about 30 degrees is achieved. The shuttle is now ready for the re-entry into the atmosphere, whose influence becomes noticeable at about 75 miles (120 km) altitude. From this point onwards the computers take care of the remaining descent to Cape Canaveral. From penetration of the atmosphere until touch-down, some 4,000 miles (6,400 km) are flown by the shuttle, the distance between Hawaii and the Cape, for it is over Hawaii that the shuttle re-enters the atmosphere. You will do this distance in thirty minutes.

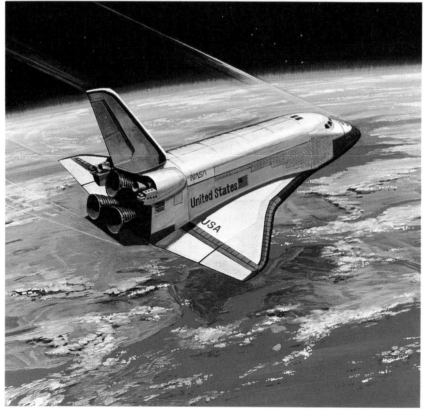

During the return to Earth the commander of a space shuttle will see an orange-coloured glow through the windows, caused by heated, electrically charged air particles around the shuttle.

Journey's end in sight.

DE-ORBIT BURN
60 mins to landing
282 km, 26,498 km/hour

TURN

COMMUNICATIONS BLACKOUT
25 mins to landing
80.5 km, 26,876 km/hour

MAXIMUM HEATING
20 mins to landing
70 km, 24,200 km/hour

EXIT COMMUNICATIONS BLACKOUT
12 mins to landing
55 km, 13,317 km/hour

APPROACH MANOEUVRES
5.5 mins to landing
25,338m, 2,735 km/hour

LANDING PHASE
86 secs to landing
4,074m, 682 km/hour

| 20.865 km | 5.459 km | 2.856 km | 885 km | 96 km | 12 km |

LANDING PHASE
86 secs to landing
12 km to runway
682 km/hour, height 4,074 m

NOSE UP (BEGINS)
32 secs to landing
3.2 km to runway
576 km/hour, height 526 m   inclination: 22 degrees

NOSE UP (ENDS)
17 secs to landing
1.079 m to runway
496 km/hour, height 41 m   pull up to 1.5 degrees

GEAR DOWN
14 secs to landing
335 m to runway
430 km/hour, height 27 m   1.5 degrees inclination

LANDING
689 m from threshold
346 km/hour

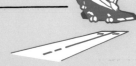

The return of the spacecraft.
The last phase is shown in the lower part of the drawing.

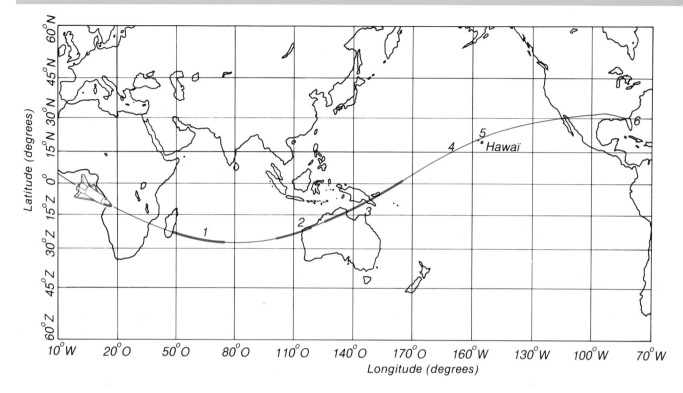

Shortly after the shuttle has entered the atmosphere the heat-shield will become white-hot, which causes the air around the shuttle to become electrically charged. This results in a loss of communication with the ground lasting sixteen minutes which is called the 'blackout' period.

The friction caused by the atmosphere has two effects: the speed of your spacecraft is reduced, and at the same time the temperature on the outside increases. In other words, forward speed energy is changed into heat energy. The nose of the spacecraft and the wing leading edges heat up to about 1,650 degrees Centigrade, and the heat-shield along the underside to 1,370 degrees Centigrade. But you will not notice the heat. At the end of the blackout period the shuttle cools down gradually, while the commander and the pilot perform various manoeuvres to adjust the speed and course to the most favourable values. At an altitude

of 14 miles (23 km) and at a speed of only 2.5 times the speed of sound (in its orbit the shuttle did actually fly 10 times faster than this), the craft is now heading for the runway, not far from the big VAB building on the Cape.

The actual approach begins at an altitude of 15,600 ft (4,720m). Apart from the voices in your headset, there is no noise: the shuttle behaves like a big cumbersome glider during its return in Earth's atmosphere. Engines no longer play a rôle.

As the 'finals' begin (the approach to the runway,) the shuttle still has a speed of 330 mph (530 km/hr). The acute angle at which the craft descends towards the runway is unusual: it is about seven times steeper than that for a passenger aircraft. But at a height of 1,700ft (520m) the nose rises and the angle becomes flatter. At 100ft (30m) 14 seconds before touch-down, the landing gear is lowered, and you land at a speed of 200 mph (320 km/hr).

So there you are then, back on

This is how you return with the shuttle from orbit:
1. Shuttle attitude is corrected
2. Firing the retro-rockets (OMS)
3. Continue rotation, nose upwards
4. Start of computer-controlled descent
5. Enter the atmosphere at about 60 miles (100 km)
6. Landing manoeuvres

Earth. A truck pulls up with the flight steps, and you can get out, feeling a little strange, no doubt, because the gravity will not make your re-acquaintance with Earth all that easy. But, without any doubt, it was all well worthwhile, and perhaps, after a few weeks, you may be longing to go back to that weightless existence in the space station, high above our densely populated planet.

The drawing below shows the last part of the return track.

The drawing opposite shows the approach and landing on runway 33, Kennedy Space Center (from NW to SE, runway 15 is used; the same runway, but now in the opposite direction, is then numbered 33. The numbers 15 and 33 stand for 150° and 330°, the courses on the magnetic compass).

A. Approximately 60 miles (100 km) from the runway threshold: altitude about 15 miles (25 km), speed Mach 2.54. Approx. 7 minutes before touch-down.

B. Approx. 29 miles (47 km) from the runway threshold: altitude about 11 miles (18 km), speed Mach 0.99. Approx. 5 minutes before touch-down.

C. Landing radar switched on. At approx. 9 miles (15 km) from the runway threshold, altitude approx. 3 miles (5 km), speed about 400 mph (640 km/hr). Approx. 2 minutes before touch-down.

D. Penultimate phase 8.7 miles (14 km) from runway threshold, altitude 15,000ft (4,600m), speed 395 mph (635 km/hr).

E. Approach: 10,800ft (3270m) above runway, 7.4 miles (11,925 m) to runway threshold, speed 372 mph (600 km/hr).

F. The nose is lowered: about 2,200ft (660m) above runway, 3.7 miles (5,930 m) to runway threshold, speed 335 mph (540 km/hr).

G. Command: 'Gear down', 215ft (65m) above runway, 1 mile (1,600 m) to runway threshold, speed 310 mph (500 km/hr).

H. Full shuttle weight on main wheels, about 1,000 yards (910m) beyond threshold, speed about 200 mph (320 km/hr).

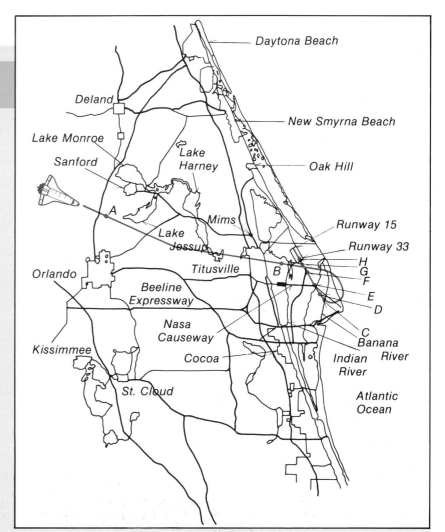

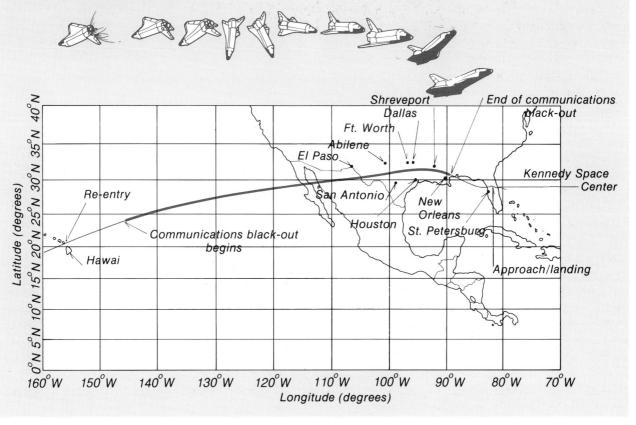

Top: The shuttle heading for Cape
Canaveral. On the right is the VAB where
the shuttle will be prepared for yet
another trip.

Top, right: Through the window in the
shuttle's entry door you can see, just
before touch-down, the place where your
journey commenced: the Kennedy Space
Center in Florida.

Centre, right: The shuttle touches down
on the Cape's runway.

Below, right: Shortly after landing you
can disembark. A NASA van (right) is
ready to take you away.

IF SOMETHING GOES WRONG . . .

Both before as well as during the launch, on the flight to the space station, and on the return, there is always a chance of something going wrong with the space shuttle. You must prepare for such an event, although the crew will assist you in every possible way.

The shuttle is fuelled on the launch platform with a large quantity of solid fuel (in both boosters) and liquid fuel (in the large external tank and in smaller tanks on board the spacecraft itself). This is a potentially dangerous situation, just like that for a passenger aircraft full of fuel.

If something threatens to go wrong before the launch (e.g. if the pressure in the tank builds up to too a high a level) you can leave the spacecraft via the access arm, the footbridge, which is rotated away from the shuttle seven minutes before lift-off. In an emergency, the arm can be put back in place quickly. You will not make use of the elevator to get down (it moves too slowly), but metal baskets which descend along tight steel cables to the ground. There are five such cages, each of which holds two people. You'll glide down very rapidly (in 35 seconds) and enter an underground bunker as quickly as you can. The 'cable-cars' can be used up to the moment of ignition of the engines. Should the three main engines of the shuttle not perform perfectly immediately after ignition, the countdown is stopped automatically. With the launch interrupted, you'll quietly await instructions from the crew, who will signify when it is safe to leave the spacecraft.

If the solid fuel boosters have been ignited and then something goes wrong, then a short flight must be made anyhow, because the boosters cannot be stopped. Nor can they be disconnected if they are functioning.

One can imagine an emergency situation in which one or two of the shuttle's main engines fail. If this occurs within the first 4 minutes and 20 seconds after the start, and the decision has to be taken to abort the flight, the commander will attempt to keep the engines operating until an altitude of 60 miles (100 km) is reached. Here the atmosphere is sufficiently thin to allow the spacecraft, still with the large external tank attached, to rotate still further backwards until the nose points in the direction of the launch site (the boosters having been jettisoned earlier). The engines, still functioning, now point in the direction of travel and quickly reduce the speed. They will then push the space shuttle towards the launching (and landing) area. At a position where the craft can glide downwards, the engines are switched off and the large tank is jettisoned; it will fall into a preselected area of the sea, away from the shipping lanes. A computer steering program (previously prepared)

carries out the critical manoeuvres until the crew can carry out a normal landing using the manual controls.

The next possibility is an emergency landing on the runway of the Naval Air Station at Rota, Spain. This is done if one main engine fails and the spacecraft is travelling too quickly for an immediate return to the Cape (as discussed earlier), and too slowly to complete an orbit around Earth.

Should the mission be aborted in the latter part of the outward trip, there will usually be sufficient thrust to take the shuttle almost into orbit. The large tank will in this case be jettisoned at the normal point, and therefore land in the planned area. The spacecraft will then have to fly once around Earth before a practically normal return and landing on an airfield in the USA.

Two astronauts trying out the emergency procedure in which they use a metal basket suspended on a cable track.

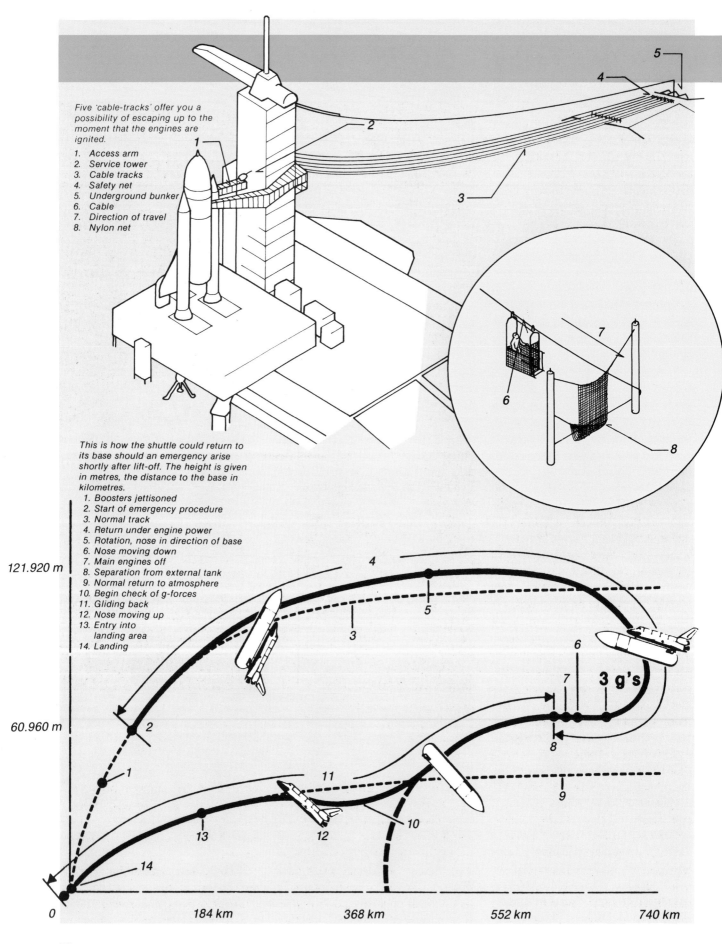

Five 'cable-tracks' offer you a
possibility of escaping up to the
moment that the engines are
ignited.

1. Access arm
2. Service tower
3. Cable tracks
4. Safety net
5. Underground bunker
6. Cable
7. Direction of travel
8. Nylon net

This is how the shuttle could return to
its base should an emergency arise
shortly after lift-off. The height is given
in metres, the distance to the base in
kilometres.

1. Boosters jettisoned
2. Start of emergency procedure
3. Normal track
4. Return under engine power
5. Rotation, nose in direction of base
6. Nose moving down
7. Main engines off
8. Separation from external tank
9. Normal return to atmosphere
10. Begin check of g-forces
11. Gliding back
12. Nose moving up
13. Entry into
    landing area
14. Landing

121.920 m

60.960 m

3 g's

0   184 km   368 km   552 km   740 km

If difficulties arise just before reaching the planned orbit around Earth, the shuttle may achieve a somewhat lower orbit, yet still be able to reach the space station. If necessary, the shuttle could use its smaller OMS engines a little longer and so still reach an acceptable orbit. Everything, therefore, depends upon the point in the launch where the problems occur, and on the nature of the problems.

But what should you do if something goes wrong while in orbit around Earth, before the shuttle has reached the space station? First, don't worry. You're better off than someone sitting in a passenger aircraft which may simply crash. At least you are in orbit around Earth; anything might happen, but you will not fall to Earth. The shuttle is a satellite which, without propulsion, can remain circling for months and years. A likely fault could be an engine failure at the moment that braking is required for the return to Earth. Even if the shuttle cannot return, there is still no call to 'abandon ship'. You wait quietly until a second shuttle arrives to take over the crew and passengers.

Other possibilities include contamination of the air supply or a leak in the cabin. In either instance it would be necessary to transport the passengers by a safe method to a quickly launched rescue craft. But there are only two space-suits on board, which are worn by the mission specialists. However, some provisions have been made for you, the passenger: there are 'space-balls', large spherical bags in which you'll sit, and which are then pressurized by means of a system which also vents the carbon dioxide breathed out. Also with you is a small portable oxygen unit, sufficient for one hour. Wrapped in such a bag you can await the rescue operation at ease. Experienced

Sitting in this life-saving ball you can be transported by an astronaut dressed in a space-suit, if you are not wearing a space-suit.

astronauts will transfer you (this is not difficult since you are weightless, remember?) to a rescue craft — a shuttle or a kind of space-tug originating from the Space Station. The transfer could take place using ropes (more or less like those used in ships), or through an airlock between two spacecraft docked together.

Only when the transfer begins will your rescue bag be disconnected from the 'umbilical cord' and you will have to depend on your own oxygen system.

It could also happen that there is fire on board the spacecraft. The more important compartments of the spacecraft have smoke detectors

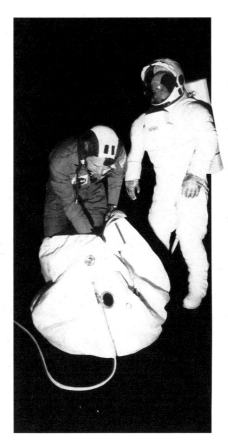

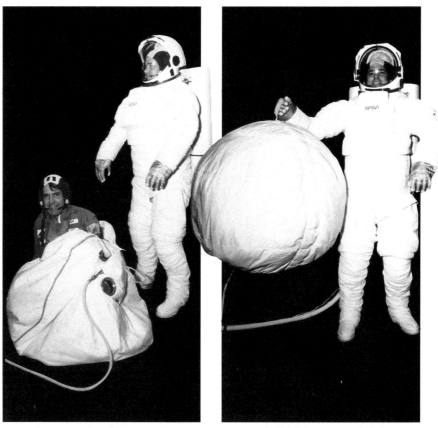

You enter the life-saving ball like this: and now you are ready to be picked up.

This is how you'll be sitting in the life-saving ball, wearing a small oxygen mask.

installed. A lamp will light and an alarm will sound as soon as a fire is detected. The crew then activate the integral fire extinguishers. Portable extinguishers are also available if required. Leave this kind of emergency action to the crew and listen to their instructions. Remember, everything is being done to make your flight as safe as possible and to deal with possible problems quickly and efficiently.

A few hours after the space shuttle has taken you into orbit around Earth, a bright star appears from afar. A short while later it proves to be not a star but a curious spider-like contraption, which reflects flashes of sunlight in your direction: the Space Station, the terminus of the shuttle service, your future domestic and working quarters in space. The Space Station orbits Earth at an altitude of 250 miles (400 km).

The distance to the enormous construction grows smaller and smaller. You'll see the large panels of solar cells, which furnish clean and cheap energy to the space 'town'; you'll see the domestic quarters, a large hangar and several space cranes. At various places are docking ports for the shuttle and other spacecraft. One of these docking ports now slowly approaches the shuttle; the last phase of the rendezvous has begun. A slight bump, and the docking is complete. You may now begin the transfer. You'll slowly float into the airlock: your initiation to a space settlement.

Versatile

The Space Station is not only the end of the line for the space shuttle; here, satellites and other spacecraft are prepared and readied for rocket-launching to higher orbits and journeys to far-away worlds. This is done using special rockets which can be used again and again. Some of these boosters can reach extremely high orbits (the so-called Orbital Transfer Vehicles), others restrict their movements to the lower orbits (Orbital Manoeuvring Vehicles). These space-tugs are serviced regularly in the Space Station, refuelled and launched again. Within the Space Station new drugs and materials are developed, and a great

deal of research is carried out. The station is presently inhabited by 16 people, but its capacity can easily be enlarged.

The space shuttle delivered all kinds of material in the early 1990s for the construction of this station in orbit around Earth, after which they were assembled to form the imposing complex which you have now boarded. In principle, there is no limit to the size of such a construction built in space.

Under normal circumstances an astronaut will remain on board the Space Station for three months. The shuttle not only ensures the transfer of people but also the supply of new machinery and basic materials, and the transport of products manufactured in the Space Station. Satellites are also supplied by the shuttle in component form. On board the station they are assembled, tested and then launched.

Thanks to the station, the time-schedule for the shuttle has become much simpler. Earlier, without a Space Station, the shuttle had to start at a definite time (the so-called launching window) in order to place a satellite in orbit at the right moment. Now, one only has to take into account the moment that the Space Station passes, and that it flies at an angle of 28.5 degrees to the Equator.

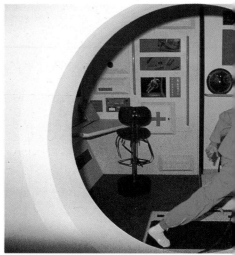

On the preceding page: Space Station, both destination and departure point. Destination for the shuttles arriving from Earth, departure point for journeys further into space, assisted by special rockets.

Free Flyers

The construction in space, which will be your home for the coming weeks or months, is only part of the total Space Station system. Also part of that system are the free-flying platforms — in the vicinity of the station or on other orbits around Earth — the so-called 'free-flyers'. Normally unmanned, such platforms could, for instance, carry large telescopes to observe the universe. These have to be vibration-free which is why the telescopes are not placed on the Space Station itself. Other platforms may carry cameras to photograph Earth, or may have tiny factories producing medicines and other products whose manufacture requires that vibration and other disturbing influences be kept to an absolute minimum. From time to time these free-flyers are visited by the shuttle crews to check the results (films, for instance) or to take away the products, to load new films, or to supply new basic materials. But this is a side issue. Let's reconnoitre the various modules of the Space Station, starting with the one in which we are to spend a major part of our time — the habitat module.

Above: The central part of the Space Station, your home in space. Two habitat modules are clearly visible.

Left: Command and control centre in Space Station; an astronaut exercising on the treadmill.

SPACE STATION

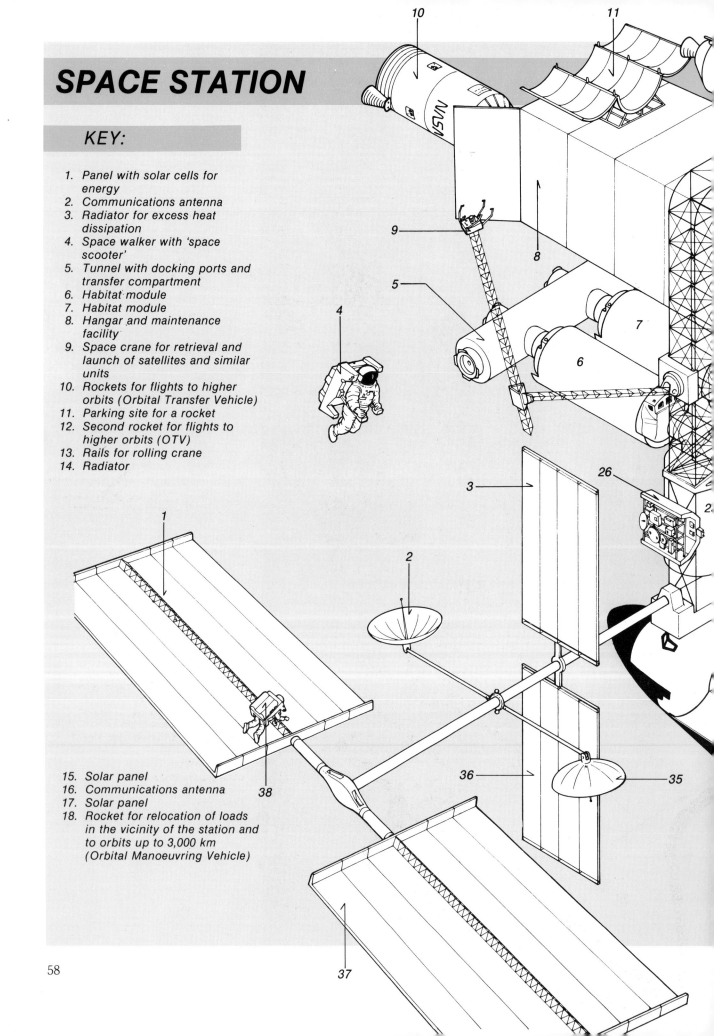

KEY:

1. Panel with solar cells for energy
2. Communications antenna
3. Radiator for excess heat dissipation
4. Space walker with 'space scooter'
5. Tunnel with docking ports and transfer compartment
6. Habitat module
7. Habitat module
8. Hangar and maintenance facility
9. Space crane for retrieval and launch of satellites and similar units
10. Rockets for flights to higher orbits (Orbital Transfer Vehicle)
11. Parking site for a rocket
12. Second rocket for flights to higher orbits (OTV)
13. Rails for rolling crane
14. Radiator

15. Solar panel
16. Communications antenna
17. Solar panel
18. Rocket for relocation of loads in the vicinity of the station and to orbits up to 3,000 km (Orbital Manoeuvring Vehicle)

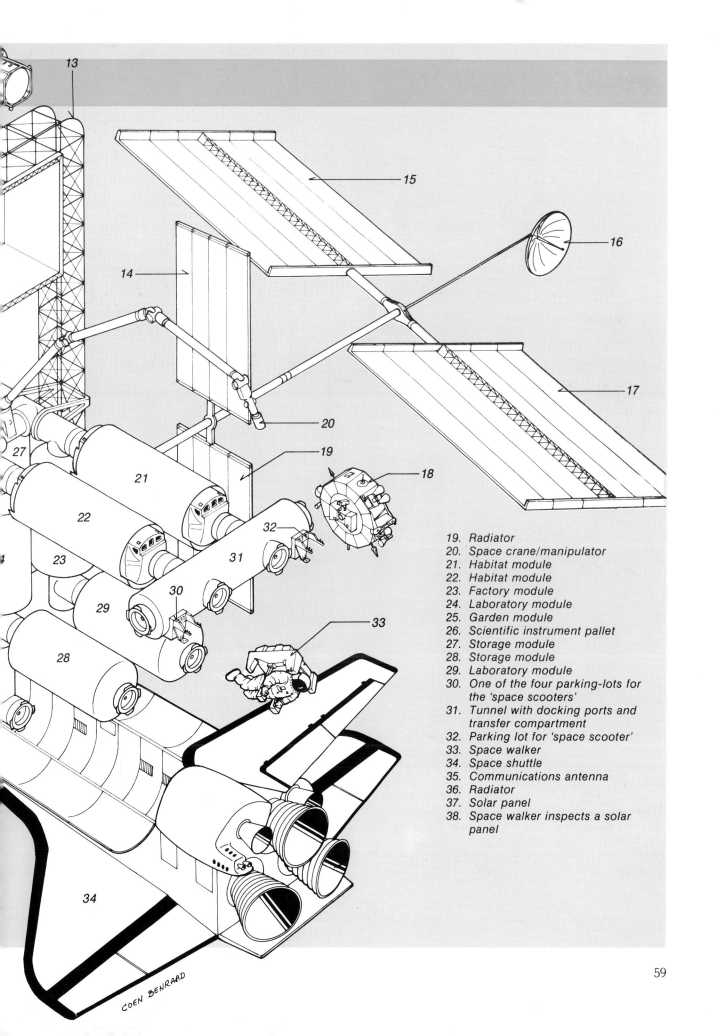

19. Radiator
20. Space crane/manipulator
21. Habitat module
22. Habitat module
23. Factory module
24. Laboratory module
25. Garden module
26. Scientific instrument pallet
27. Storage module
28. Storage module
29. Laboratory module
30. One of the four parking-lots for the 'space scooters'
31. Tunnel with docking ports and transfer compartment
32. Parking lot for 'space scooter'
33. Space walker
34. Space shuttle
35. Communications antenna
36. Radiator
37. Solar panel
38. Space walker inspects a solar panel

COEN BENRAAD

The arrangement inside one of the four-man habitat modules of the Space Station.

1. Command and control centre
2. Toilet, washing facilities and shower
3. Dressing room
4. Space for physical exercise and recreation (including home-trainer)
5. Bedroom with writing desk, light, air-conditioning and TV
6. Ditto
7. Observation deck for looking at Earth, the Moon and the Universe (with storage space for personal property, cameras, telescopes/binoculars and scientific instruments)
8. Small bedroom
9. Ditto
10. Dining room (the table has built-in waste disposal unit)
11. Kitchen (with store for water and food, a small refrigerator and an oven)
12. Wardrobe for space-suits
13. Entry to airlock for space walks, including docking device

Dimensions:
Length: 50ft (14.95 m)
Diameter: 14ft (4.30 m)

The domestic module is shown here cut in half for clarity. From left to right: Command and control centre, toilet, washing facility and shower, dressing room, kitchen, wardrobe for space-suits, entry to airlock with docking device. Beneath the floor: All systems are housed here; power supply, atmosphere regeneration, control, etc.

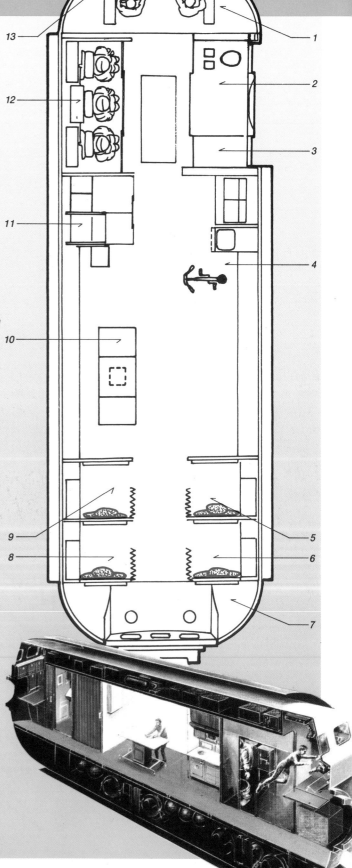

From the space shuttle, you have entered, via the airlock, one of four identical habitat modules in which you will be working and living during the coming weeks — perhaps even for as long as three months. The drawing and the artist's impression will give you an idea what your quarters look like; a place where everything has been adapted to life without weight. It will take some getting used to during the first few days, but then you'll discover that practically everything has been taken care of. For instance, there is a sitting-room (even though you hardly ever 'sit') — where you can carry out physical exercises at the same time — a dining area, a kitchen, a laundry and toilet room, and four small bedrooms. Along the front and rear of the more than 48ft (14m) long, luxurious, cosmic 'bungalow' are large windows which offer a fine view of Earth, the universe and the complicated structure of the Space Station. At the end is the command cabin, where the controls are installed for the habitat modules and other parts of the Space Station. At the opposite end you'll find the observation deck, where you'll spend many pleasant hours.

From here you can look at Earth and take photographs. There are telescopes of different types, through which you can bring nearer all the things of interest to you — for example, the Moon and the planets.

Experience teaches that people in space never get tired of looking at our blue planet, which offers ever-changing views while the station performs its laps: every 90 minutes a complete journey around the world. Jules Verne would have been surprised! Each habitat module is a very complex bit of machinery which ensures that you can live in comfort. Above the ceiling as well as under the floors are machines to clean the air and the water for re-use. And there is also storage space for those things not wanted every day.

Above: The living room with the kitchen.

Right: A view of the living room in the factory mock-up on Earth.

DRESS FOR WORK AND LEISURE

Neither in the space shuttle nor in the Space Station do you have to wear a space-suit. The standard dress consists of a two-piece cobalt-blue flight suit: a jacket and a pair of trousers. The cloth is soft cotton, specially treated to be fire-resistant. Underneath the suit one normally wears a dark blue T-shirt, usually adorned with an emblem of the Space Station; the flight suit's jacket bears a similar emblem. All clothing comes from NASA's clothing store and is not 'made to measure'. The jacket is chosen according to arm and chest measurements, the pair of trousers according to waist and leg measurements.

The clothing, at first sight, is not indicative of the fact that it has been made to be as safe as possible. The suit is comfortable, but not sloppy. Loose clothing is quite bothersome when weightless: you could catch it, for example, on a switch — with who knows what results. The jacket has baggy shoulder-pieces, improving mobility when manipulating things. The suit has a dozen pockets in it, in which all kinds of useful things are stored so that they don't start to float all around you and drift out of sight. On departure from Earth these pockets contain felt-tipped pens, pressure ballpoints (which will also write under weightless conditions), pencils, small books containing data, notebooks, sunglasses, a Swiss combination knife with all kinds of handy tools built-in, and, finally, a small pair of surgical scissors (for opening food-packs). A small torch has not been forgotten either. The jacket is closed by a zip-fastener, but the pockets lack buttons or zippers; velcro has been used instead, which allows the pockets to be opened and closed quickly and easily. The dress is the same for men and women. You will receive one set of underclothes per day (including a bra for the women, although breasts require less support when weightless).

It stands to reason that socks and light shoes have not been forgotten (years ago Cosmonaut Valery Sewastyanov complained about holes in his socks due to the continuous pushing he had to do — they had not given him any shoes).

For each flight everyone receives a pair of gloves, although they are seldom required.

This is the standard equipment. If you want, they'll issue you with some shorts: in the Space Station it is very comfortable and warm, and it is pleasant to wear shorts every now and then. So equipped, you can go for a 'swim' to your heart's content in your temporary abode — at high level. But beware: it takes some getting used to!

Dress:

Underwear	:	1 set daily
Shoes	:	1 pair per flight
Jacket	:	1 pair per flight
Trousers	:	1 pair per flight
Shirt	:	1 every 3 days
Gloves	:	1 pair per flight
Bra (females)	:	1 daily

Personal equipment:

Felt-tip pens and ballpoints	:	as required
Propelling pencils	:	as required
Swiss pocket-knife	:	1 per flight
Pair of scissors	:	1 per flight
Watch	:	1 per flight
Sunglasses	:	1 per flight
Sleeping mask and earplugs	:	1 set per flight

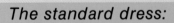
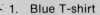

The standard dress:

1. Blue T-shirt
2. Jacket
3. Pair of trousers

The standard dress for space-dwellers

If more informal wear is required, a pair of shorts may be worn instead (see opposite page).

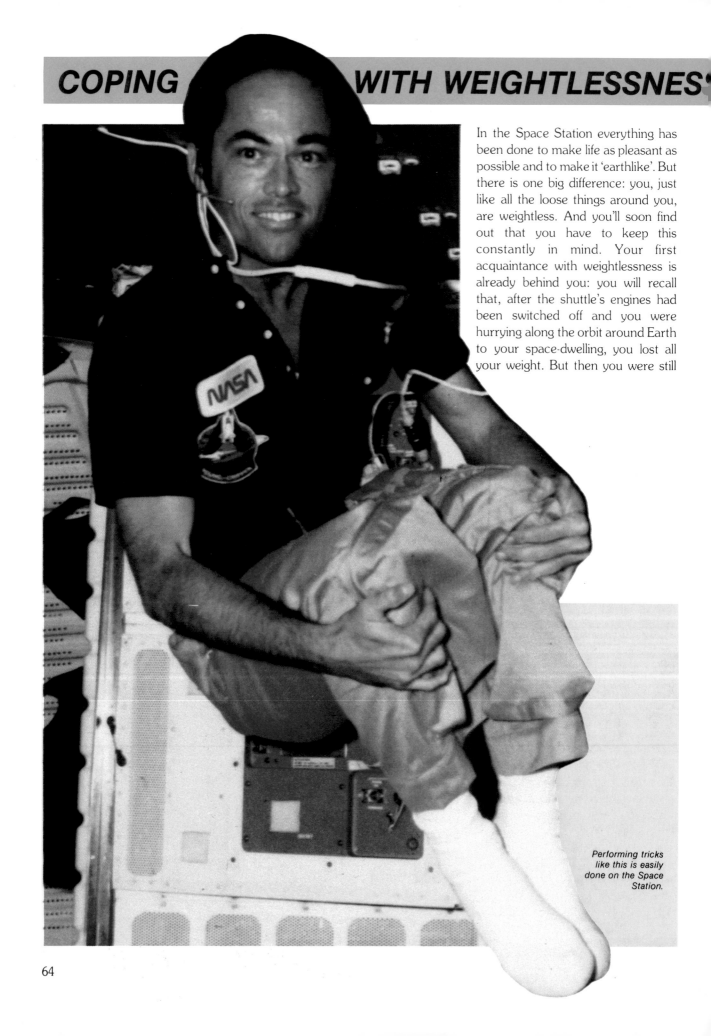

In the Space Station everything has been done to make life as pleasant as possible and to make it 'earthlike'. But there is one big difference: you, just like all the loose things around you, are weightless. And you'll soon find out that you have to keep this constantly in mind. Your first acquaintance with weightlessness is already behind you: you will recall that, after the shuttle's engines had been switched off and you were hurrying along the orbit around Earth to your space-dwelling, you lost all your weight. But then you were still

Performing tricks like this is easily done on the Space Station.

strapped good and proper to your seat and so hardly noticed the absence of gravity. But once you have transferred and floated into your space abode, you will be confronted with it to the full.

Now you too must rough it a bit, because more than half of all space travellers experience unpleasant sensations during their first few days in space. The first few minutes you'll not notice much. Usually, the first irksome phenomena to appear are dizziness, excessive perspiration and even vomiting. This is an indication that your balance-sensor, of which the most important parts are in the internal canal of the ear, has been seriously upset. Under normal conditions you'll know where is up and where is down. In space you'll lose that sense. You'll have to learn to eliminate your balance-sensor gradually and completely, and to rely only on your eyes. Even just seeing someone upside down (you will instinctively assume that you yourself are standing up or floating) can give you a feeling of nausea. Pills used against motion-sickness for car, ship or aircraft do not help here. It is now essential to make as few abrupt head movements as possible, otherwise it's guaranteed that you'll be sick. And it could reach the stage where you wish you'd never left good old Mother Earth.

With one finger only . . .

To assist you as much as possible, the station has been made to show that there is definitely a 'down' and an 'up'. The floors are dark in colour; the ceilings and walls are lighter. You'll feel better if you move calmly, holding on to the hand-grips installed in many places in the station, and ensure that you are upright as indicated by your senses. You'll notice that your arms

Playing the 'he-man' is easy – provided your object is weightless.

This kind of photograph can only be taken if normal gravity is absent.

When weightless, movement is very easy, but it takes a little while to get used to it.

A star made up of space-dwellers in a laboratory module.

and legs are behaving differently from on Earth. They are, unless you react consciously, not straight but bent at the joints. Your arms have a tendency to float in front of you, unless you purposely keep them near you.

After a few days you have attuned yourself nicely and now you can do some more experiments. Simply by using one finger you'll slowly drift in a chosen direction. And then it appears that weightlessness also has its pleasing features, because everything is very easy, and it takes little energy to move yourself from one spot to another. You can also caper around; rotate very fast around your axis, or perform loops which would have been impossible on Earth.

Fat Head

It is now time for you to retire to your own personal bedroom and take a look in the mirror. Here you'll find to your amazement that all the lines and wrinkles in your face (if you had any) have disappeared, and also that your head has become fatter, more 'pumped up'. And your eyes seem smaller than they used to.

This is due to the blood flowing more to your head than is the case on Earth, where the gravity pulls the blood downwards. As a result your waistline has decreased, so that you'll have to tighten your trouser belt.

Your feet have also shrunk a little, so you'll have to tighten your shoelaces. There is less blood in the lower body, while the total volume of the blood has decreased because the quantity of blood in your head acts as a signal to your body that you have too much blood. Your body has also secreted more salt than would be the case on Earth.

These are the effects that will be visible within a short period. But generally you'll grow one to two inches taller: the discs between your vertebrae are expanding a little because the vertical pressure has decreased. In time your muscular mass will reduce, and some calcium will have disappeared from your bones. These phenomena are not really alarming, but if you intend to live in a Space Station for several weeks or months then it is essential to spend one or two hours each day at physical exercises, so keeping your muscles at a reasonable level, and to work out regularly with varying loads. You must also make regular use of a piece of apparatus which ensures that the blood is pulled down to your lower body. This is a kind of cylinder into which you'll go up to your waist. An elastic tape ensures the air-tight sealing of the cylinder and then the pump sucks some air from the cylinder. This causes a vacuum, thereby causing extra blood to be 'sucked' into the arteries in the lower body. All these precautions will

ensure that you'll be strong enough to withstand the gravity on Earth and that you will not faint immediately upon return.

Parabolic flight

The situation is rather like that of people who have been cloistered in their beds for weeks on end, and then suddenly have to get up. If you rise too quickly after a single night's sleep it can happen that you feel dizzy. After a long period of weightlessness this effect is, of course, much stronger. (By the way, this gives you an indication of how to prepare yourself on Earth for your weightless existence in space: by much lying in bed (!) and by elevating the bottom of your bed a little so that your head is slightly lower than your feet. In this way the blood circulation to your head is improved and your body can begin to adjust itself to a situation which you'll meet while in orbit around Earth).

You can also form a pretty good idea about weightlessness with a so-called 'parabolic flight' on board an aircraft. This method is used for all professional astronauts. It happens like this: first, the aircraft flies at cruising height and then dives, as if into an imaginary valley. Then it pulls up sharply. At this point, not only the aircraft but everything and everybody in it becomes three times heavier than normal: 3g, as it is called. Then the aircraft climbs over an imaginary mountain and the throttles are now closed. During this climb the aircraft and its passengers experience exactly the right upward accleration to counteract the gravitational force.

Getting acquainted with weightlessness is possible in an aircraft making parabolic tracks. Each time this happens, it will provide between 15 and 30 seconds of weightlessness. However, it is not always funny . . .

Result: the aircraft and passengers become weightless. Once over the top of the mountain, the aircraft can't keep on diving of course, the engines come alive again and pull the aircraft out of the valley again. In the jargon, parabolas have been performed. Each time, you can be weightless from a quarter to half a minute, and so you may get some idea of what weightlessness entails.

Adaptation

Once in orbit around Earth, there is still quite a bit of adapting to do. During the whole time, you must bear in mind that gravity is absent. You can only stay in your seat if you have fastened yourself by means of a belt. A pen, a fork or any kind of tool cannot just be put down unless you fasten it to something. If you break a slice of bread, the crumbs will not drop onto your plate, as would be the case on Earth, they will start floating around and could put out of action some piece of apparatus, or get into your lungs. An attempt to put salt on your steak would only result in the salt flying in all directions. If you spill water it will start to drift about in droplets. And when these droplets touch the wall they spread over the surface as a thin film of moisture, very difficult to remove.

On Earth you subconsciously make much use of gravity. A paper weight is useless to you in space. You cannot have a photograph of your beloved relations on the wardrobe in your bedroom. If you want to open a panel, as you used to do on Earth, and you bend forward towards the panel, there is a fair chance that you'll keep on making endless somersaults. So you'll first fetch a portable handgrip equipped with suction-pads. This handgrip you'll attach close to the

Two female astronauts create a weightless work of abstract art.

panel, and then you can open it normally. So you must be prepared for everything you do in space to cost you more in energy than would be the case on Earth.

Except hanging around, of course.

If you have a job to do outside the station or want to hang around there purely for your own pleasure, then you'll need a space-suit. This can be further complemented by a one-man rocket-chair, also called a 'space scooter'.

PUTTING ON YOUR SPACESUIT

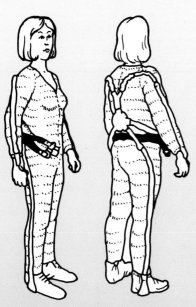

1. Underwear and ventilation

2. Pull up trousers

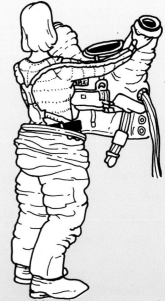

3. Prepare top half

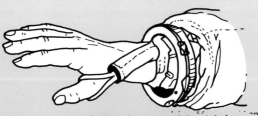

5. Note: loop around thumb **before** sliding into top half

4. With arms up, slide into top half

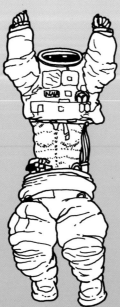

6. Connect cooling-water connection from underwear to survival pack coupling

7. Connect the two halves of the space-suit

Your space-suit

1. Water-cooled underwear keeps your body at the correct temperature
2. Umbilical cord comprising communications and electricity, water and oxygen lines. Connections between underwear and space-suit
3. Electrical connections to the survival pack
4. Control box (carried on the chest)

5. EVA visor, which is slid over the helmet during space walks
6. One-piece helmet made of plastic (polycarbonate)
7. Arm of space-suit with movable shoulder and elbow

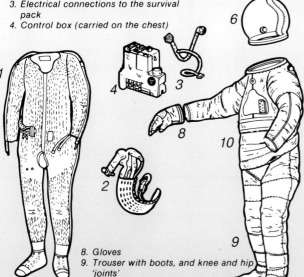
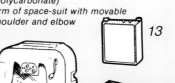

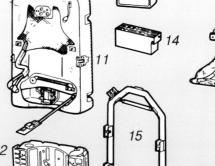

8. Gloves
9. Trouser with boots, and knee and hip 'joints'
10. Stiff top-half of space-suit
11. Survival pack
12. Reserve tank for thirty minutes' extra oxygen

13. Filter box for air purification
14. Rechargeable battery
15. Bracket for space-suit on the wall of the Space Station

16. Cap with microphone and headphones
17. Drinking bag carried in top half of space-suit. The drinking tube ends inside the helmet
18. Urine-collector with drainage installation

Space-suit control box

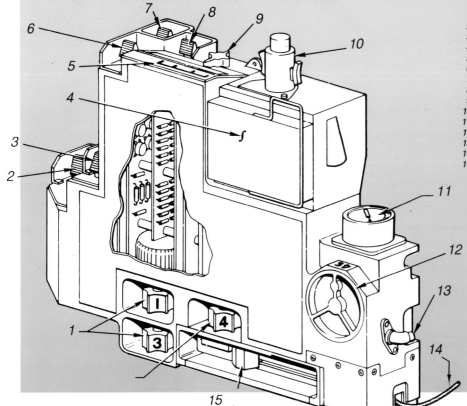

1. Communications volume control
2. Press-to-talk switch
3. Ventilator switch
4. Cover over connections
5. Display
6. Safety-system switch
7. Power switch
8. Water-supply switch
9. Communications selector switch
10. Cleansing valve
11. Pressure connection
12. Temperature control
13. Oxygen connection to survival pack
14. Cable for oxygen control
15. Oxygen control

Your space-suit

Your space-suit is in fact a one-man spaceship. The survival pack on your back not only contains sufficient oxygen and water (you can drink a little via a tube in your helmet) to last approximately seven hours, but also a ventilator to circulate the oxygen in your suit, and a pump to pump the cooling water around it. It also has a container with lithium hydroxide and a carbon filter to absorb the carbon dioxide and vapours you exhale. Then there is a cooling installation to cool the water, and the batteries used to supply electricity to the apparatus. The suit is made up of several layers. The basic layer is a rubber sack filled with oxygen, and provides a pressure of 0.25 kg per square centimetre, i.e. about a quarter of the pressure on Earth. This low pressure is necessary to prevent the suit from becoming stiff, but it is high enough to prevent your blood from boiling. Over the rubber layer is a layer of dacron-polyester to maintain the shape of the suit and to prevent excessive expansion. On top of this are several layers of fireproof fabric and flexible metal, for protection against temperature variation and radiation. The temperatures on the outside of your suit may vary from minus 157 to plus 121 degrees Centigrade.

Standard sizes

Inside your domestic module and in other places in the station you'll find a number of space-suits in standard sizes. The time when space-suits were made to measure (e.g. for trips to the Moon) is long gone. Each suit consists of two major components: a top part and a bottom part. They are connected together by means of rings around your waist. The torso is made of aluminium and therefore is not flexible, but the sleeves are connected to it through rotating rings. The shoulders and elbows have bellows to allow movement of the arms to be as smooth as possible. Without this the bending of an arm would be like bending a firmly inflated balloon, because the space-suits are pressurized since there is no atmosphere outside the station. A survival pack is attached to the top half of the suit, which can supply you with oxygen and water for up to seven hours: fifteen minutes for checking the suit after dressing, six hours for activities outside the Space Station (EVA — Extra Vehicular Activity), fifteen minutes to take off the suit, and thirty minutes to spare. In addition there is a separate small, emergency oxygen tank for thirty minutes. The survival pack can be regenerated and refilled from the Space Station's systems.

A small box is attached to your chest on which you can check how much oxygen and battery power remain. Here too are the inescapable warning lights, which will light up if something is not working properly. The lower half of the suit consists of flexible trousers with joints which bend adequately. Under this space-suit you'll wear a special overall, as underwear. Through this run tubes along which (via the survival pack) water is pumped, and surplus heat is drained. There are also hoses which

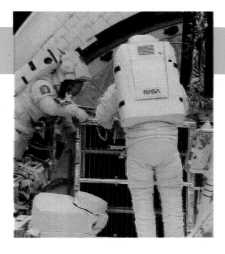

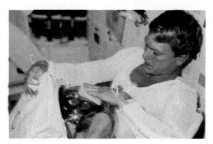

take oxygen to many places around your body, so ensuring the necessary ventilation. Of course, the space-suit is only complete once the pressure gloves, the bubble-helmet (a one-piece item) and a quickly attachable and detachable visor have been added. The gloves, available in 15 different sizes, have special rubber fingertips which give you some sense of touch when using tools, etc. The visor on your helmet protects the eyes against micro-meteorites and the ultraviolet rays emitted by the sun. Do close the visor when you go outside, otherwise you will surely burn if you stay too long in the unfiltered sun.

In five minutes

You can put on your space-suit within five minutes without any help. It will be necessary to try this a few times and to get somebody to explain everything to you. First, you put on the long underwear plus the reservoir for collecting the urine. Next you pull on the trousers and then you'll slide into the top half of the suit. You must make certain that the many connections on the top and bottom halves connect properly, so that the water and oxygen can flow from top to bottom and vice versa. Then you'll put on the helmet and pull on your gloves. Ensure that the fasteners are closed properly, because even the smallest leakage could prove fatal. After a thorough check you will get into the airlock and close the inner panel. Press a button and the air will flow away slowly. Meanwhile, keep on checking your indicators to determine that your suit is functioning properly. If everything is all right, and the pressure inside the airlock has been reduced to zero, you must fasten a double safety-rope which will prevent you drifting away helplessly into space. Now open the outer panel and you can get out. Make sure that you have a torch on you because you will be in darkness for about three-quarters of an hour during each orbit. It is preferable for you to have an experienced companion for your first walk in space. This can safeguard you against much unpleasantness!

'SPACE-SCOOTER'

When, dressed in a space-suit, you arrive outside via the airlock, you will remain tied to the Space Station with two safety-ropes. If this were not so, you would slowly drift away from the station. In space there is an almost absolute vacuum. Waving your arms and legs around only results in you performing all kinds of gyrations around your own centre of gravity. Nothing else will happen. If you are to work close to an airlock of the Space Station, you can move **one** safety-rope at a time, using the handrails which have been installed everywhere. In some locations you'll even find some foot-supports which will facilitate working there. If you have to move over greater distances around the Space Station, or some distance away from the station, then, of course, you'll be using the 'space scooter'. You will exit from the airlock and move, using the handrails (but still attached to the safety-rope), to the 'parking-lot'. Here, you'll attach yourself to the scooter, unfasten the safety-rope, and fly into space.

Now don't get the idea that you can tear through space on the space scooter, as is done in science fiction books. This is not possible, because in space it is very difficult to estimate distances. Acceleration and deceleration have to be carried out very gradually. Theoretically, if you are to travel one hundred metres from the airlock to a satellite floating near the station, then you'll accelerate for fifty metres. Then you'll begin to decelerate at exactly the same rate until, having arrived at the satellite, you are practically motionless. However, in practice it does not work out that way, as you have only your eyes to estimate speed and distance. But there is a rule of thumb: if you are to cover a hundred metres your best speed will be **one** metre per second.

Which means that it will take 100 seconds to cover that distance. In practice it takes a little longer. So you do it like this: accelerate to a speed of 1 metre per second, which is 3.6 kilometres per hour, or slightly less than normal walking pace. When you think you have reached that speed, just float close to the target and then decelerate. In this way you can transport yourself quickly, the risks are at a minimum and very little fuel is used.

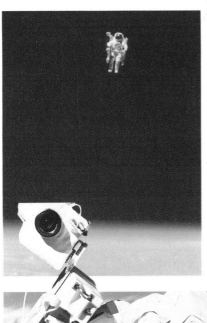

Top: Space mechanics building a large platform in space; for example, a solar power station. Of course, both carry MMUs on their backs.

Right: This is how you'll inspect a satellite, close to your base in space.

MMU 'Space Scooter'

With the MMU (Manned Manoeuvring Unit), commonly called the 'space scooter', you can move in space outside the Space Station to your heart's content. The MMU can move along the X, Y and Z-axes, while rolling, pitching and yawing are possible, which means that rotational movement is possible in three dimensions perpendicular to each other. These movements are made possible by 24 jets, fed by gaseous nitrogen.

Commands for these movements are given via two controls. Near the right hand is a rotatable knob for controlling rotation, and near the left is a stick for moving forward, to the rear, to the left, to the right, upwards and downwards. These two therefore control all movements. Near the right hand is another knob which, when depressed, maintains your current exact position.

There are also some switches on the arms of the MMU for switching on and off the position controls and the electronics. Switches to prepare the MMU for use are mounted on the protruding parts of the MMU near the astronaut's shoulders. The batteries are stored in the back-rest.

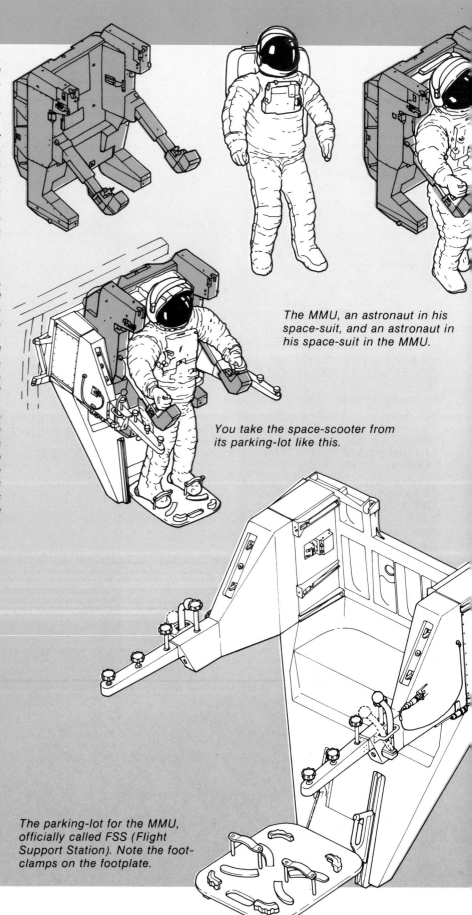

The MMU, an astronaut in his space-suit, and an astronaut in his space-suit in the MMU.

You take the space-scooter from its parking-lot like this.

The parking-lot for the MMU, officially called FSS (Flight Support Station). Note the foot-clamps on the footplate.

The MMU's arms are adjustable, length-wise, and can be lowered to 102 degrees below the horizontal. During displacements, the arms are 30 degrees below the horizontal; and when the MMU is parked, 102 degrees below the horizontal (folded backwards). Mechanical couplers, to keep the MMU in place, are located in the shoulder and hip components of the MMU. The couplers connect to the large survival pack on the back of the space-suit. To be released from the MMU, the astronaut pulls one or both rings in the shoulders and the survival pack is freed.

The MMU is 4ft 1in (125.4 cm) high, 2ft 7 ins (82.7 cm) wide and 4ft (120.9 cm) deep (with the armrests in the 30 degree position). It weighs 330 lb (149.7 kg), with the tank filled with 26 lb (11.9 kg) of nitrogen. The total weight of an astronaut and a space scooter lies between about 646 an 740 lb (293 and 336 kg), depending, of course, on the astronaut's weight.

All vital systems in the MMU are duplicated, so that if one of the systems fails there is another available. Normally both sets are used, but in an emergency one of them can be switched off. However, fuel can still be drawn from the two nitrogen tanks in this case.

The MMU gets its electrical supply from two silver-zinc batteries. As well as for the controls and steering, the power is also used to feed the signalling lights, making the MMU visible a long distance away from the Space Station.

It will be obvious that each of the two batteries serves its own system. Each system comprises twelve small jets. Each nitrogen tank, when full, contains 13 lb (5.9 kg) of nitrogen. Of course, the pilot can check at any time how much gas is available. An MMU can be used for about seven hours before being due for re-charging and servicing.

The MMU is parked in a Flight Support Station (FSS), on the outside of the Space Station and in the freight compartment of the

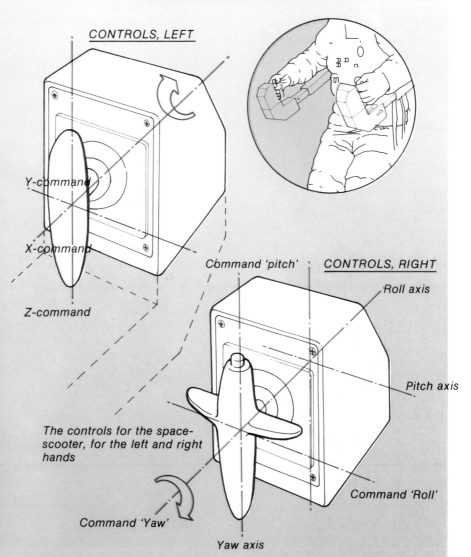

CONTROLS, LEFT

Y-command

X-command

Z-command

Command 'pitch'

CONTROLS, RIGHT

Roll axis

Pitch axis

The controls for the space-scooter, for the left and right hands

Command 'Roll'

Command 'Yaw'

Yaw axis

space shuttle. Every FSS has a footplate with straps, under which the astronaut can anchor his feet when picking up and returning the space scooter. The footplate is adjustable to suit the astronaut's height. Each arm of the FSS has four mushroom-shaped knobs for the astronaut to hold when getting on or off. On the left and right-hand sides are handgrips for freeing the MMU and for fastening down.

The FSS also serves as a filling station for the MMU. From the Space Station (or the shuttle) the recharging of the gas and the batteries is monitored. The MMU is also kept at the current temperature through the FSS.

EATING AND DRINKING IN SPACE

Do people living in space only eat 'mash' from tubes? No, and they haven't done this for some time. In the early days of space travel, cosmonauts and astronauts did indeed squeeze their tubes until empty, drank via a tube and ate pre-cooked delicacies. In those days the spacecraft were rather small and, what is more, they were afraid that crumbs or liquid droplets would start floating through the cabin and cause faults in the spacecraft's equipment. That period is definitely behind us. In the Space Station you will discover that the meals are tasty and varied: more than one hundred different foods are available. Naturally, there is some standardization, and the food is vacuum-packed, dehydrated and deep-frozen. Many foods have been dried to save weight and space: cornflakes, vegetable soup, spaghetti, scrambled eggs, bananas, pears, strawberries and shrimps. Also, twenty different drinks like tea and coffee are available in powdered form. Real orange juice, grapefruit juice or whole milk are not included. If you add water to deyhdrated orange juice or grapefruit powder you'll find loose grains in the water, while dried whole milk starts to form lumps when water has been added. Therefore, synthetic

A ready-made bite floating towards a hungry astronaut.

juices are used, and skimmed milk. Other food is supplied to the Space Station in natural form, or pre-cooked and packed in tins, irradiated or treated by some other method. Examples of 'natural food' include crackers, walnut biscuits, peanut butter, nuts, sweets, chewing gum, tinned food (for instance, tuna) and fruits in thick syrup.

Foods subjected to radiation (to prevent formation of fungi and bacteria) include, for instance, meat and bread. The variety of food is so extensive that it will be more than a week before you are served with the same meal again.

The diet has been planned especially for you, the traveller in space, in order to counteract the loss of certain minerals, which tends to occur under weightless conditions. Sodium, calcium, nitrogen and other trace elements influence our muscle power and the strength of our bones, not to mention the effect upon mental processes like the ability to concentrate properly. So plenty of these minerals are put into space food. Breakfasts, lunches and dinners are packed as meals, in containers stored in the kitchen. Once every 16 days it will be your turn to prepare the meal for the 16 crew members. You take off the outer wrapping and place those tins and boxes that require heating in the oven. At the right-hand side of the galley is a punching machine which you can use to insert hot or cold water into the packs of dehydrated food. If necessary, you can do the same with the food placed in the oven, which has reached a temperature of 82 degrees Centigrade.

When everything is ready you put the food into the four trays, which, as on normal aircraft, have recessed sections, in which the food packs are held in place by magnets (tin) or velcro tape. The tray is fastened to the table by clamps. There are magnets on your tray to hold your knife, fork and spoon. Before you start eating you must cut every plastic pack with a special pair of scissors, so that you can get at the food. Most of the food you'll be eating is encased in a thick gravy or jelly-like substance so that it sticks to your spoon or fork — on all sides. This takes a little getting used to, because you'll tend to take a larger bite than you do normally. The danger of food floating away is not

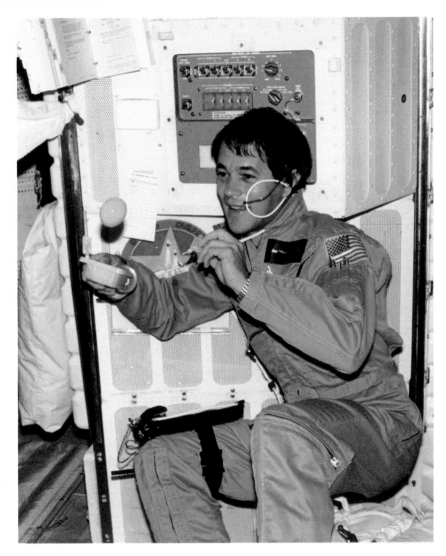

For your breakfast: a ball of orange juice.

very great. Should something float away from your cutlery, it is no problem as it can be easily caught because it will stay near you. Easier still is to open your mouth and take the piece of potato or the weightless sphere of soup directly into your mouth. This is easier than catching a peanut with your mouth open — on Earth, that is.

Sandwiches with dressing can be eaten without problem. Fruit juices and alcoholic drinks (only on public holidays) are in closed plastic packs into which a straw is pushed. Nuts and snacks are available in abundance in the habitat module's galley. To give you an idea of what's cooking in space, there follows a normal 3,000 calorie menu for three days.

Day 1

Breakfast
Peaches
Meat paste
Scrambled eggs
Bran
Chocolate milk
Orange juice

Lunch
Sausages
Turkey
Bread
Banana
Almond candy bar
Apple juice

Dinner
Shrimp cocktail
Steak
Rice
Broccoli
Fruit cocktail
Caramel pudding
Grapefruit juice

Day 2

Breakfast
Apple purée
Dehydrated meat
Muesli
Bread roll
Chocolate milk
Orange or grapefruit
juice

Lunch
Corned beef
Asparagus
Bread
Pears
Peanuts
Lemonade

Dinner
Meat with barbecue sauce
Cauliflower cheese
Runner beans with
mushrooms
Lemon pudding
Walnut biscuits
Chocolate milk

Day 3

Breakfast
Dried peaches
Sausage
Scrambled eggs
Cornflakes
Chocolate milk
Orange or pineapple Juice

Lunch
Ham
Cheese spread
Bread
Beans and broccoli
Pineapple purée
Shortcake
Cashew nuts
Tea with lemon and sugar

Dinner
Mushroom soup
Smoked turkey
Mixed vegetables
Vanilla pudding/strawberries
Tropical punch

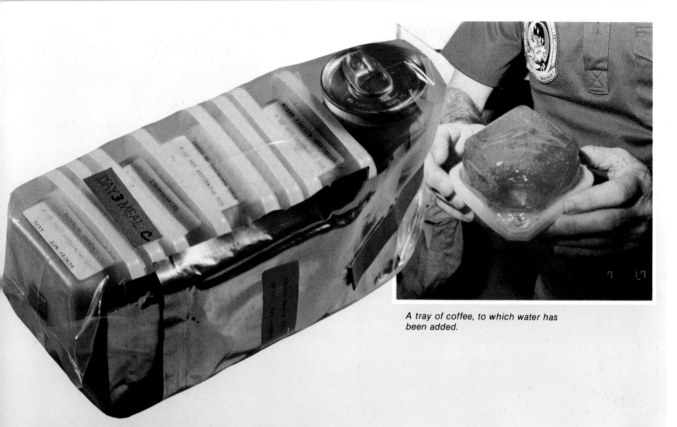

A tray of coffee, to which water has been added.

In the kitchen you punch a hole into a box of food and add (as required) hot or cold water through a hollow needle.

Condiments: pepper, salt, barbecue sauce, ketchup, mayonnaise, mustard.
Drinks are available as follows: tea, coffee, apple juice, chocolate milk, lemonade and various fruit juices.

SLEEPING IN SPACE

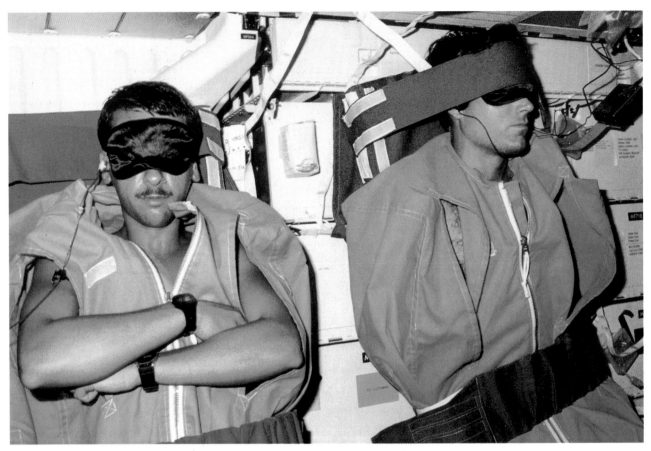

Just like eating, sleeping in space is a little different from on Earth, due to you being weightless. To grab a few weightless winks does not, strictly speaking, require a bed. A few space dwellers prefer to seek a quiet corner somewhere in the Space Station and close their eyes. If you do this you must find a place to anchor yourself to, a strap, for instance. If you don't, then, in your sleep and propelled by your own breath, you'll start roaming around the interior, risking a painful awakening when you hit something. The first astronauts, of course, slept in the seats in which they launched because there was no other place. But they quickly discovered that their arms started to float during the sleep, which involved the risk of hitting a knob somewhere with their hands. So they put their hands under their safety-belts. There is ample room in our Space Station, and hence small single bedrooms. To an outsider it may appear that you are standing rather than sleeping, but in space the position is completely immaterial. The bedrooms are arranged to take up as little space as possible within the habitat module. Each bedroom has a bed 6ft (1.8m) long and 2ft 6ins (0.75m) wide. It is, in fact, no more than a plank covered with a soft material to which a sleeping bag has been attached. The whole bag is perforated for ventilation.

On Earth, where your weight is normal, your body sinks into the mattress. In the Space Station, you'll hardly feel the hard board beneath you. But you'll have the pleasant illusion that you're lying on something. And that feeling stimulates sleep in most humans.

Above: Two astronauts sleeping in space.

Right: 'Dropping off' is a term that loses its meaning in space.

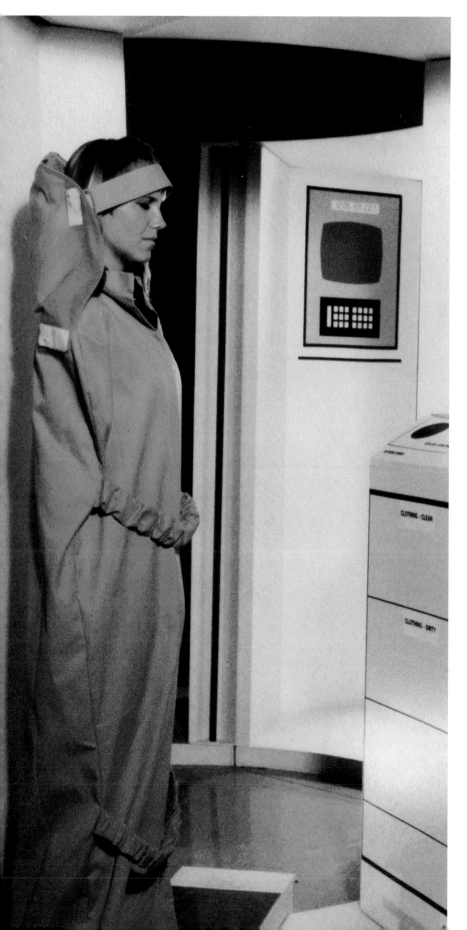

If you feel sleepy (every day you'll witness sixteen sunsets, so you can't depend on those!), you take off your clothes and carefully store them in the drawers and cupboards in your bedroom, so that you'll not be rudely awakened by a shoe hitting your nose. You can sleep in your underwear or put on pyjamas. Open the zip-fastener and crawl into your sleeping bag, after which you close the zipper again. Tighten the two belts across your chest so that you are held firmly in your sleeping bag. In each bedroom a mask and ear-muffs are available in case you wish to sleep undisturbed. Ensure that light cannot enter your bedroom, because that is not recommended for your rest. As stated earlier, every hour and a half the Space Station orbits Earth and the sun rises every hour and a half, to set three-quarters of an hour later. Before going to sleep adjust the ventilation and switch off the light (the switches are close to your head), after which you slide your arms under one of the belts to stop them floating in front of your face.

Good night!

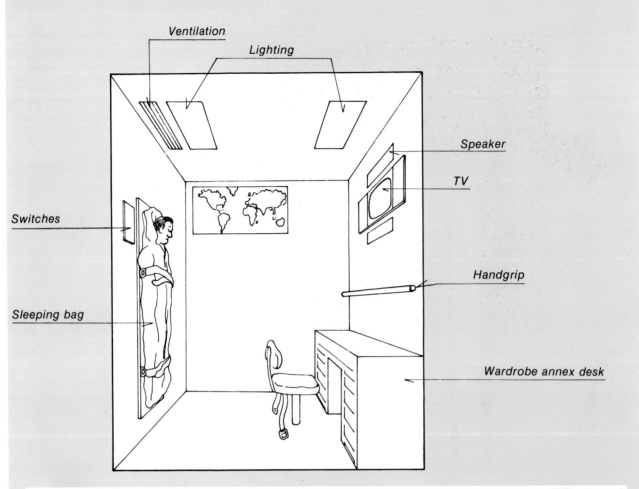

Ventilation

Lighting

Speaker

TV

Switches

Handgrip

Sleeping bag

Wardrobe annex desk

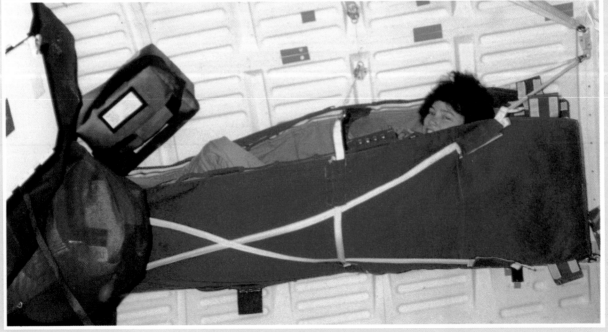

Drawing, left: Your bedroom in space.

Below, left: No need for beds; sleeping bags can be attached anywhere.

Below, right: If you don't slip your arms under a strap or into a suit, they will start to float about while you sleep.

WASHING, SHAVING, MAKE-UP

The habitat module in the Space Station is equipped with every comfort in respect of your personal hygiene. For example, you will find a shower in a small, lockable cubicle. Should you want to take a shower, you must first undress in the dressing space then float into the shower and anchor your feet firmly under the straps on the floor. Otherwise you may finish upside-down in the shower cubicle — unless that's what you intended. Close the door to prevent droplets of water spreading out from the cubicle.

Water droplets

Use the hand-shower and the liquid soap in the normal way. There are holes in the floor of the cubicle through which the water is drained off under suction to prevent the weightless droplets of water from quickly filling the cubicle. For obvious reasons you cannot take a bath on the Space Station. You would quickly be surrounded by drops of water and have great difficulty in 'keeping your head above water'. Not until we have artificial gravity for our Space Station (by slowly rotating a wheel-shaped station) will a normal bath be possible. However, a shower on board a Space Station is a great improvement compared with the shuttle, where a wash with a washing glove is all that's possible.

When you have finished taking a shower, wait until all the water has been drained from the cubicle and then get out to dry and dress yourself. Even then it is advisable to make use of the various footstraps because

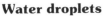

On Earth, a hairdresser would have some difficulty in achieving this hair-do, but while weightless it happens by itself – providing you have sufficient hair.

This space traveller has few difficulties with his hair!

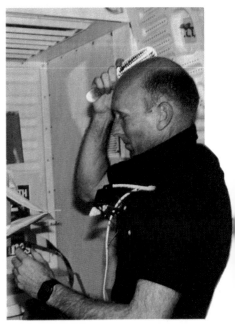

otherwise the fight with your clothes could have little or no result. Of course, you can wash your face and hands without taking a shower. You just use a washing glove. While weightless you require very little water, because the water adheres very well indeed to your skin. Drying yourself is no different from on Earth.

Wash-basin

The wash-basin looks different. Again, this has to do with the absence of gravity. First select hot or cold or somewhere in-between. Turn on the tap and push your hands through the two sleeves into the transparent plastic sphere, as shown in the drawing. The wash-basin is also connected to a vacuum pump to stop the water getting out.

For each person in the habitat module, NASA will supply a personal hygiene kit: a folding bag containing a toothbrush, toothpaste, toothpick, nailclippers, soap, a comb, a brush, make-up and lipstick for women, skin lotion, and a deodorant stick. For male space travellers there is a tube of shaving cream and a safety razor or an electrical or clockwork shaver. You will receive seven washing gloves and three towels per week. Naturally there is a small laundry machine in the domestic module. For waste disposal there are containers which are emptied into space or which go with the shuttle as return freight. Garbage bags are ejected under power to prevent the Space Station being surrounded by a cloud of them, as happened to the first Russian Salyut. Even from Earth — using a telescope — it could be seen that the little bright star moving overhead (the Salyut) was surrounded by minute pinpoints — garbage bags put overboard.

Unbelievable, but true!

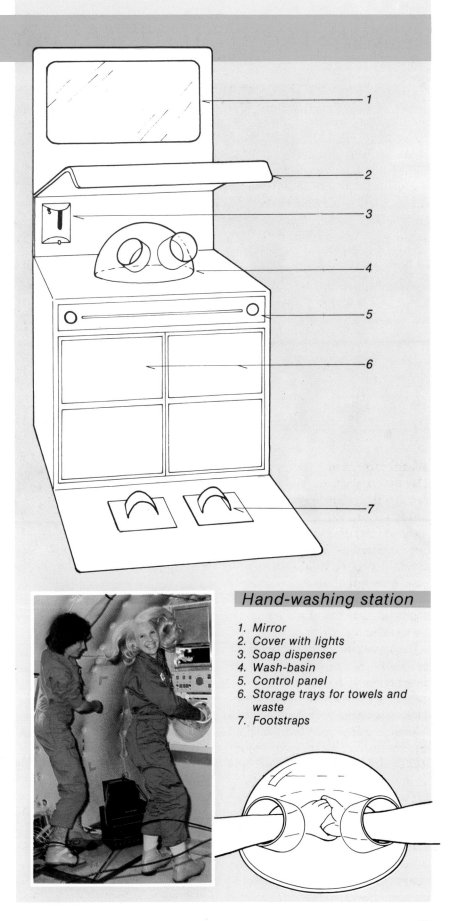

Hand-washing station

1. Mirror
2. Cover with lights
3. Soap dispenser
4. Wash-basin
5. Control panel
6. Storage trays for towels and waste
7. Footstraps

Far right: How to wash your hands in the wash-basin.

Right: Two astronauts under training try to wash their hands on board an aircraft in which weightlessness is created for a short period.

USING THE TOILET IN SPACE

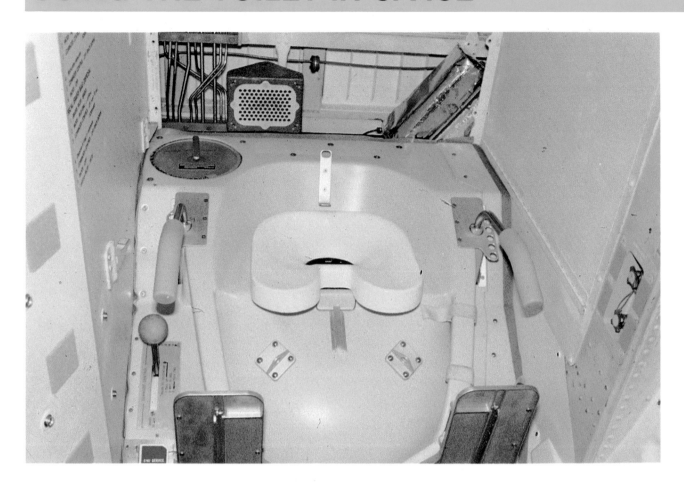

This delicate subject must be discussed thoroughly so as to protect you against unpleasant surprises because everything you release is weightless, and this doesn't stop in 'the smallest room'.

The habitat module has a real toilet to the left of the wash-basin. It is unnecessary to tell you what to do first, but once you have reached that stage, place your feet firmly in the footclamps on the toilet floor. Do not forget to fasten your seat belt — and put it on really tight. You can also hold on to the handgrips. It is important that your behind forms a seal around the rim, closing off the toilet bowl, because this toilet is not flushed by water but operated by air. If this is allowed to escape through an opening then the system functions less well, as has been proven in practice.

The air is stirred by a fast-rotating centrifuge which you switch on by moving the lever on your right to its furthest forward position. At this moment several vanes begin to rotate, which together produce oscillation of the air. This oscillation breaks up the firm parts of the faeces and flings them forcibly against the inside of the toilet bowl where they form a fixed, odourless film. The airstream ensures that the materials strike the fast-rotating vanes. Urine of both male and female users is collected in a funnel under light suction: a pump collects all liquids in a tank which is periodically emptied into space. The integral container of the toilet can easily be exchanged when its maximum capacity has been reached. When you move the handle to 'off' after using the toilet and the

inner lid of the toilet is closed, the contents will be subjected to the vacuum of space, drying the faeces very quickly.

This toilet is of the same type as that on board the shuttle. A big improvement compared with the early period of space travel, when astronauts had to answer the call of nature by putting their faeces in a kind of plastic hot-water bottle into which some pills had been dropped which performed the function of a chemical toilet. Bags of this nature were rolled up after use and placed in the compartment which had earlier contained the sandwiches: the logical sequence!

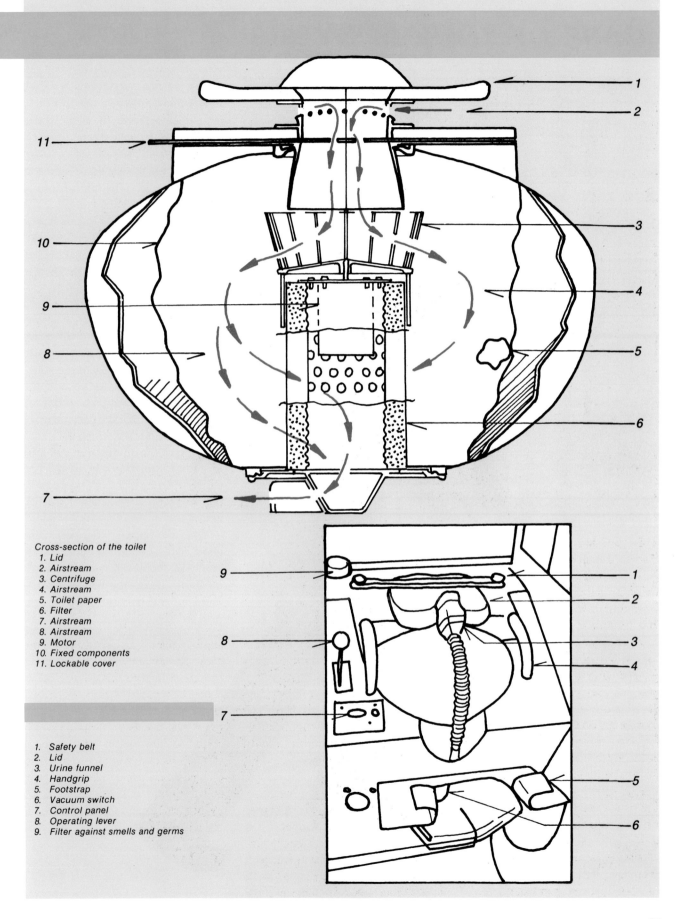

Cross-section of the toilet
1. Lid
2. Airstream
3. Centrifuge
4. Airstream
5. Toilet paper
6. Filter
7. Airstream
8. Airstream
9. Motor
10. Fixed components
11. Lockable cover

1. Safety belt
2. Lid
3. Urine funnel
4. Handgrip
5. Footstrap
6. Vacuum switch
7. Control panel
8. Operating lever
9. Filter against smells and germs

CARE OF YOUR BODY

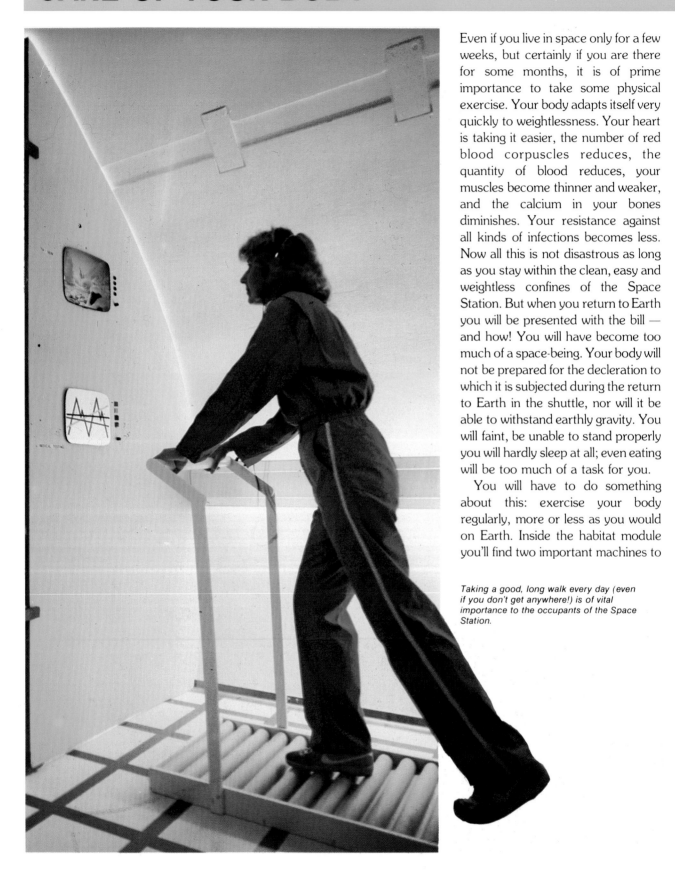

Even if you live in space only for a few weeks, but certainly if you are there for some months, it is of prime importance to take some physical exercise. Your body adapts itself very quickly to weightlessness. Your heart is taking it easier, the number of red blood corpuscles reduces, the quantity of blood reduces, your muscles become thinner and weaker, and the calcium in your bones diminishes. Your resistance against all kinds of infections becomes less. Now all this is not disastrous as long as you stay within the clean, easy and weightless confines of the Space Station. But when you return to Earth you will be presented with the bill — and how! You will have become too much of a space-being. Your body will not be prepared for the decleration to which it is subjected during the return to Earth in the shuttle, nor will it be able to withstand earthly gravity. You will faint, be unable to stand properly you will hardly sleep at all; even eating will be too much of a task for you.

You will have to do something about this: exercise your body regularly, more or less as you would on Earth. Inside the habitat module you'll find two important machines to

Taking a good, long walk every day (even if you don't get anywhere!) is of vital importance to the occupants of the Space Station.

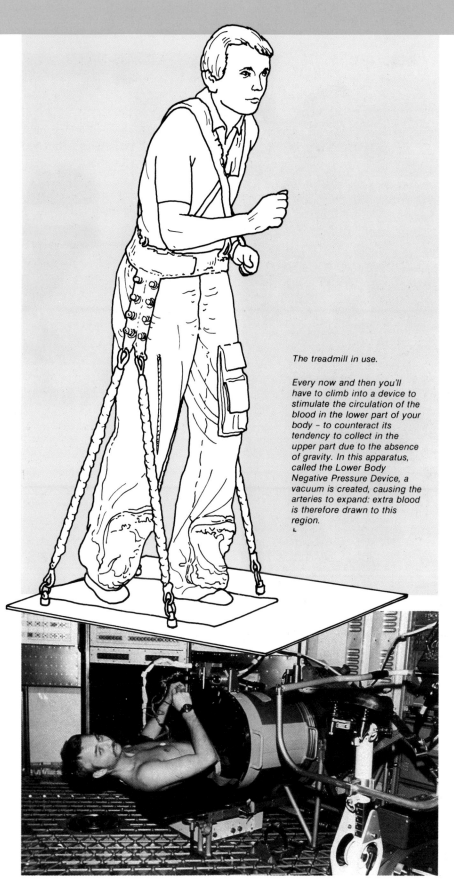

The treadmill in use.

Every now and then you'll have to climb into a device to stimulate the circulation of the blood in the lower part of your body – to counteract its tendency to collect in the upper part due to the absence of gravity. In this apparatus, called the Lower Body Negative Pressure Device, a vacuum is created, causing the arteries to expand: extra blood is therefore drawn to this region.

do this: a 'treadmill' and an exercise bicycle. The treadmill, which you must use for at least half an hour daily, enables you to take a walk without going anywhere. Spring-loaded straps, which are fastened to a belt around your waist and, at the other end, to the floor, will hold you against the treadmill at any speed. The tension of the straps can be adjusted to make the going more or less difficult.

A second important training gadget is the exercise cycle. A home-trainer in fact, on which you can cycle with different loads applied. You can not only adjust the load on the instrument panel but also check your performance. The habitat module also has some springs, expanders and similar devices to help you to maintain your fitness. Weights or dumb-bells are not available, as they are obviously of no use. If you are staying in space for some months, you must exercise for at least two hours every day. Make sure that the air ventilation is on 'high' during the exercises, to ensure fast removal of perspiration, otherwise you will have a thick layer of sweat on your body, which will block the pores. This moisture can then only be removed with a kind of vacuum cleaner, because being weightless it does not run down but sticks firmly to the skin. So keep an eye on the air-conditioning when you are exercising.

By the way, if you are a slave to nicotine you'd better remain on Earth, because on board the Space Station smoking is absolutely forbidden, for safety reasons.

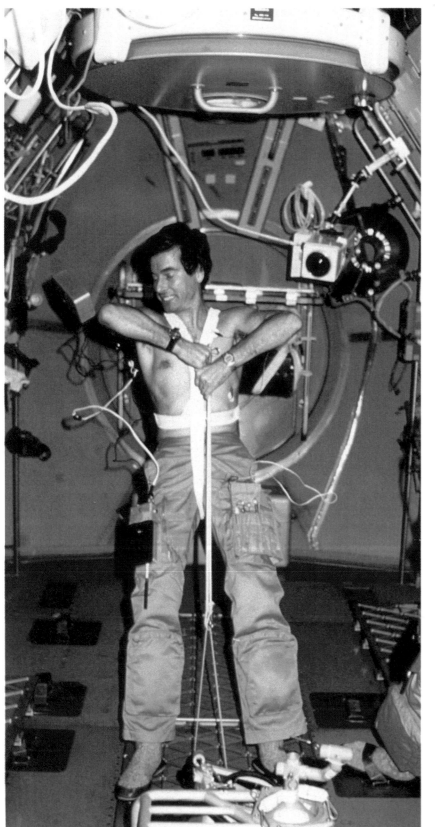

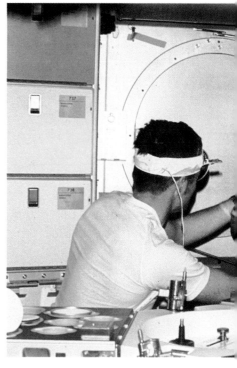

Space doctor

On board the Space Station there will be a doctor who can give you medicine should this be necessary — for instance, during the first few days on board, you may feel nauseous because your sense of balance is disturbed. The space doctor can also treat your teeth, and even remove your appendix if necessary. It would be rather expensive to transport a person back to Earth for such simple surgery. It is reasonable to assume that adequate medical apparatus is on hand in the station, like a cardiograph, a stethoscope, blood-pressure monitor, a defibrillator (for re-starting the heart) and surgical instruments.

Top: Your weightlessness makes it easy for the dentist to inspect your teeth.

Left: The daily work-out on board the Space Station.

SAFETY

To check your weight, ordinary scales would be useless, as you have no weight whilst in orbit around Earth. Therefore a mass-meter is used. This has a swinging seat: the more power required to swing the seat, the greater the mass.

Safety above all. This is the motto, not only during your outward-bound and homeward-bound journeys in the space shuttle but also of course during your lengthy stay on board the Space Station. It orbits in an almost perfect vacuum around Earth. At this altitude there is no atmosphere to serve as protective blanket. And, in this vacuum, radiation dangerous to the human body (from ultra-violet radiation from the sun to cosmic rays) as well as meteorites will penetrate, while the absence of air and air pressure is of course the biggest threat to Man.

Inside the comfortable Space Station you will notice little or nothing of all this. The air inside has the same composition and pressure as on Earth. And the thick, layered walls of the Space Station have been constructed to offer maximum protection against radiation and micro-meteorites. The danger of meteorites – small rocks flying through space at speeds of up to 44 miles (70 km) per second – was over-exaggerated in the past. In science fiction books, space travellers are still targets for meteorites. But it will rarely happen that a Space Station is hit by a meteorite large enough to make a hole so that the air escapes from the station. Most meteorites are minute dust particles against which the Space Station is adequately armoured. A large meteorite (for example, one a few inches in diameter or even larger) would, of course, render your habitat module airless in one strike. But the risk of being hit by such a rock is no greater than your house on Earth being hit by such a cosmic rock, because the atmosphere covering the Earth does not really offer any protection against these kinds of whoppers.

Should the module you are occupying really be struck by a meteorite which makes a hole, then we can assume that it will be a small one. This means that the air pressure will go down slowly. You will hear and see the warning, and without any panic will leave for another module. The modules are connected via hatches or sometimes airlocks, so you can always withdraw safely.

More dangerous of course is a short-circuit or overheating somewhere, causing a fire. Fortunately, a fire in space will not spread very quickly. This is because the gases produced by the fire itself will stay around the fire, so that fresh oxygen cannot get at it. Quite different from on Earth, where hot air is lighter than cold air and rises. In other words, a fire in space will quite quickly smother at its source, especially if the air circulation is switched off promptly.

All modules of the Space Station have integral fire detectors and fixed fire extinguishers. Of course, portable fire extinguishers are also available. The commander of the Space Station will inform you about the safety procedures on your arrival. Also available in all modules where people live or work are oxygen masks which you can put on if, for instance, the air has been polluted due to a fire.

RELAXING IN SPACE

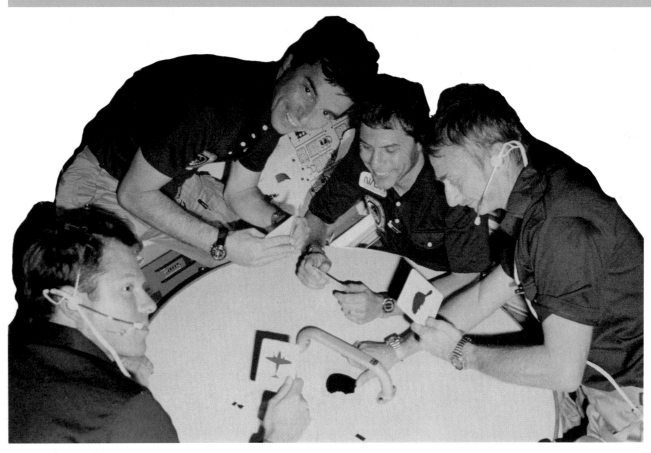

You can relax on board the Space Station in many ways, actively as well as passively. Weightlessness now offers all types of unexpected possibilities, although you will have to learn how to handle them. As far as active relaxation is concerned, you can make use of the home-trainers to stay fit. This aspect has already been discussed in the pages dealing with physical exercise and medical care.

Ball games

An obvious game is one with a ball. You can play this easily even within the somewhat limited confines of the Space Station's modules. Of course, balls will bounce back off the walls of the station just as they would on Earth. The trajectories described by the balls are, however, only determined by the actions of the contestants, the elasticity of the ball and the natural law that the angle of incidence is equal to the angle of reflection. When you throw a ball its track is not influenced by gravity, and this is something you will have to learn to account for. Incidentally, catching a ball is not at all easy either. You will still have an instinctive impulse to jump at an oncoming ball, but then there is a fair chance that you'll sail past without catching it. But, again, you must play to learn. And, after all, winning is not everything. It could well end up being a question of who moves most, the player or the ball. Ball games in space often present a somewhat dizzy spectacle of bodies tumbling all over the place, trying vainly to get hold of a ball. In view of the limited space it is a great advantage that balls do not have to move fast in weightless conditions. On Earth they must, otherwise they

Having a game of cards in space is not so easy as it seems. Cards will not stay on the table, and throwing them down is quite out of order. So elastic bands are fitted in the centre of the table to keep the cards in place.

would not get very far. In space you can push a ball very slowly. It is possible to push the first ball, to be followed quickly by a second one, with the object of hitting the first ball. In this way a kind of three dimensional game of billiards is created which can, of course, be played with more than two balls. And you can also put spin on a ball by giving it a flick as you push it. If you find it more comfortable, you can anchor yourself via foot-straps on the floor. Table tennis is perfectly possible and it has the advantage that it requires little space.

Darts

A very popular game on board the Space Station is darts, throwing little arrows. This game also requires some getting used to because on Earth you involuntarily consider gravity when throwing. You aim slightly higher because you know that gravity will pull down the dart during its flight. In space this does not apply. A thrown dart keeps on flying straight until it hits the target. And again, you don't have to throw hard. You can even give the dart a tiny push, so that it floats toward the bull's-eye extremely slowly. During a game of darts everything is made a lot easier if you fix your feet, but it could also become pleasantly complicated if you throw while unattached. You will have to remember that immediately after the throw you will rotate around your own point of gravity. This applies to every movement you make, if you are not anchored.

Chess

Other games on the Space Station are hardly different from on Earth. Chess and checkers (draughts) will be played, using magnetic pieces and magnetic boards, as otherwise the game could rapidly become very confusing. This applies also to all other games in which pieces, tokens, discs or what have you, are used. The use of dice, required for many games, is impossible in the Space Station. A thrown dice is guaranteed to go all over the place, and never really 'lies' anywhere. Hence the use of a dice-disk, a kind of wheel of fortune to which you give a firm push and then wait to see what the score is. Use is also made of a mini computer, which after pressing a button shows a random number, based purely on Lady Luck.

Computer games are also 'in' on board the Space Station, just like on

Earth. Of course, you can spend your leisure time doing all kinds of small tests which have to do with the absence of gravity. A number of things you learned at school do not work out in the Space Station — though sometimes they work much better. A pair of scales does not function in space, neither does a spring-balance. Determining the mass of something can only be achieved by subjecting it to a known force. As the accleration decreases, the mass becomes greater. And vice versa, of course. And now, **you** try to think of new tests and games which are only possible under weightless conditions.

Space wrestling

A bout of wrestling — hopefully friendly — is as crazy a business on board the Space Station as is trying to

Looking from your window in the Space Station you will see ever-changing panoramas. Here two space-walkers are inspecting a communications satellite. One of them has anchored himself to the end of the manipulator arm.

Having a snack . . .

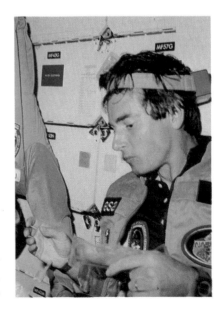

chase someone. Once you have pushed yourself away you cannot do anything about it until you touch a wall, or something else, and you then push yourself away again. Once you have got hold of somebody you start to tumble around together; but if you release your hold for just a moment, you will float apart again. You could agree, before beginning, that you win the bout if you manage to hold your opponent against a wall, floor or ceiling for ten seconds. But this can only be achieved if you can secure yourself firmly. Any possible amorous escapades in space are naturally also subject to this kind of restriction. The old-fashioned sleeping-bag on the wall could well be the most obvious place, although someone with a vivid imagination could well envisage other, new possibilities!

And if, on completing your adventures in space, you wish to share these with your family and friends on Earth, then you can make video recordings or take beautiful photographs. Shots taken through the windows of your 'spacious' dwelling are beautiful too!

Space offers a rich source of inspiration to draughtsmen and painters. And drawing and painting will go well, as long as you take care to attach yourself and your things firmly. And then we have passive relaxation. You can watch television and play video recordings. Via the TV you can also speak to your dear ones way down on far-away Earth. Or listen to music, or just read a book.

Through the window you see one of your companions taking a trip through space.

Inset, top right: The clouds in a low pressure area show a spiral-shaped pattern. The shuttle, with the freight-doors opened, is suspended at the top of this photograph.

Top right: And this is a constant subject for discussion: a dawn or sunset, which occurs every three-quarters of an hour.

Right: Talking to his son . . .

Left: Earth remains fascinating. This is what the Horn of Africa (Somalia) looks like from a height of several hundreds of kilometres.

In the space laboratory scientific research is conducted into new materials.

Top right: You can take some beautiful shots with a camera from a spacecraft. At several hundreds of kilometres above Earth, you can see the Straits of Gibraltar with Spain on the left. Morocco on the right.

Below, right: The human body is also the subject of research in the high-level laboratory.

If you have a scientific background you can carry out investigations on board the space station. The station has been equipped with a special laboratory module. The Space Station offers unique opportunities for research for three reasons: it rotates very fast around Earth; it is high above Earth's atmosphere; and there is weightlessness.

Taking everything into consideration, all the work carried out on the space station is bound up with one or more of these special conditions.

Speed (once around Earth every one and a half hours) means that large parts of our planet can be observed in a short period of time and photographed. The fact that the space station is above Earth's atmosphere (at about 250 miles, or 400km) means that all the radiation coming from the Universe can be captured. Most of these rays (for instance, ultra-violet and X-rays from the Sun and the stars) do not reach the surface of Earth, because they are absorbed by its atmosphere.

Finally, weightlessness offers unique possibilities for investigating those processes for which the gravity always prevailing on Earth is important — in a negative or a positive sense.

Special Possibilities

A large number of scientific experiments can be conducted in the laboratory module: the cataloging of the Earth's environment; the study of the influence of weightlessness on human beings; for testing parts of future satellites; the study of the Universe; and the investigation of all types of new materials, like new alloys, special crystals and medicines.

For new alloys, weightlessness offers special opportunities. It's like this: if you try to melt two different metals and then to mix them to form an alloy, this will only be successful if the densities of the two metals are similar. On Earth, the lighter metal always floats on top, as oil floats on water. But in the Space Station, where there is no gravity, the differences in density ('specific gravity' it is called) have no influence. It is even possible to mix molten metal

with air. After cooling, a kind of aerated bar is formed: very light, yet very strong. Similar new metal alloys do not have to be used only in space; once they have solidified they can be transported to Earth. Research takes place in the laboratory module, but large scale production will be carried out in the factory module (described elsewhere). It is here that special medicines are also produced.

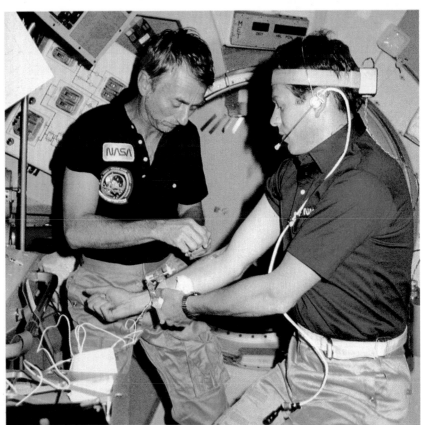

Life

In the laboratory module you can carry out research on the growth of living things (plants, animals and humans) during prolonged weightlessness, because all life on Earth developed there over millions of years while subjected to the force of gravity. By removing all gravitational force, all kinds of properties, and possibilities of living beings become more apparent.

Much of the research in the framework of the space station project does not, however, take place in the laboratory module but on platforms attached to the Space Station, or somewhere in its vicinity, or even in quite different orbits. For instance, there is a research platform in orbit over the Earth's poles, which — among other things — is very suitable for photographing Earth, which itself rotates under the orbit, so presenting every point on Earth twice a day to the lenses of the cameras and to the instruments.

The Nile-delta (left, top), the Sinai desert, the Gulf of Aqaba and the Gulf of Suez.

Right: A magnificent view of the Choi-da-mu basin in Eastern China.

4

MADE IN SPACE

Due to the absence of gravity, space is an ideal place for making a number of products which cannot be manufactured, or are very difficult to produce on Earth, in the presence of the ever-dominating gravity. The manufacture of these types of product takes place in a special factory module. The basic ingredients are supplied by the space shuttle, which takes the resulting products back to Earth again. The items manufactured here mainly comprise new alloys, crystals and new medicines.

Some medicines are certainly much better made in space than on Earth. Like, for instance, drugs for diabetes, interferon (an excellent way to counteract a large number of infections), drugs for diseases of the blood and for treatment of serious burns. These mainly utilize substances based on living matter of human or animal origin. These materials can be separated from the biological basic material by applying electrical currents. The process is called 'electrophoresis'. This is already being done on Earth, but here the inevitable gravity spoils things. Gravity is much stronger than the subtle electrical currents applied during the process of separation. Result: from a large quantity of base material only a little of the substance can be prepared, and even then it is not pure. And, to top it all, it is expensive. In space, gravity is absent and so the electrical currents function better. Through all this, large quantities of the substances become available, which facilitates commercial production.

The first introductory experiments were carried out in the space shuttle at a time when there were no space stations at all. It then appeared that, given the same quantity of base material, about 700 times more of the substances could be obtained than on Earth. And five times purer, too. Needless to say, the production of such materials means a profitable business in orbit around Earth, and hundreds of millions of people have the benefit of these drugs, which formerly were only available in small quantities.

We have already mentioned new alloys. And they have already been absorbed into our society. The same applies to very large crystals in which enormous quantities of information can be stored, and which would make the old-fashioned tape-recorder appear antique. A recorder without any moving parts is perfectly possible. A sturdy crystal has, of course, a

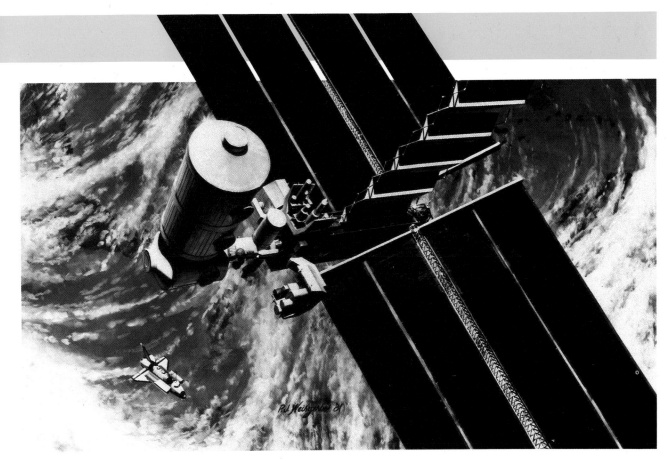

bigger volume than a flat tape. The information is stored with the aid of laser rays. These crystals are the successors to the 'talking chips' which we used in a great variety of machines in the eighties, but which could not 'remember' more than just a few words. Large crystals are also the ideal basic material for making high-quality chips. The electronics industry depends increasingly upon electronic products from space which, labelled 'Made in Space', are already on the market. Inside the factory module these and many others can be made — and you can watch. And again it is the space shuttle which takes care of the supply of raw materials and transport of the products.

Above: A future space factory, functioning automatically, and visited only occasionally by astronauts for maintenance purposes, and for the supply of raw materials and the collecting of products.

Opposite: The space shuttle is used for supplying materials and collection of products manufactured in space.

Top: A free-flying factory in space in orbit around Earth.

5

GARAGE IN SPACE

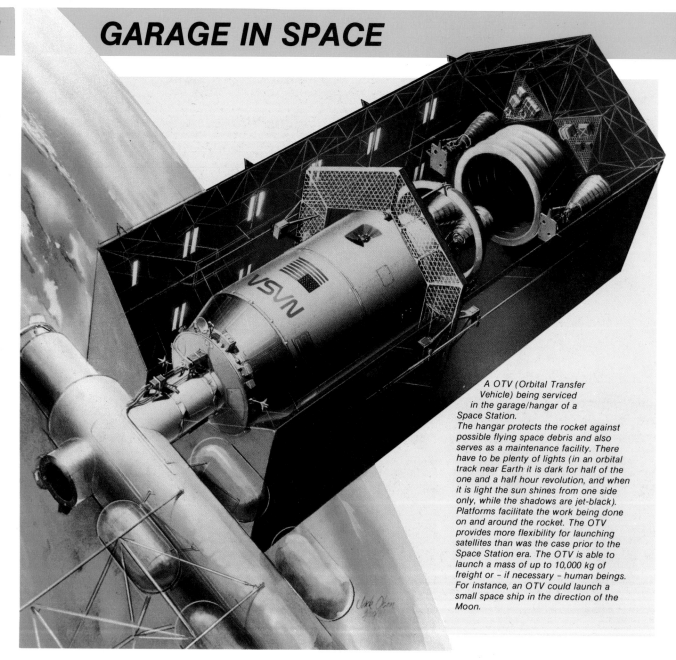

A OTV (Orbital Transfer Vehicle) being serviced in the garage/hangar of a Space Station.
The hangar protects the rocket against possible flying space debris and also serves as a maintenance facility. There have to be plenty of lights (in an orbital track near Earth it is dark for half of the one and a half hour revolution, and when it is light the sun shines from one side only, while the shadows are jet-black). Platforms facilitate the work being done on and around the rocket. The OTV provides more flexibility for launching satellites than was the case prior to the Space Station era. The OTV is able to launch a mass of up to 10,000 kg of freight or – if necessary – human beings. For instance, an OTV could launch a small space ship in the direction of the Moon.

If you have sufficient skill you can occupy yourself in the Space Station with the assembly, testing, launch and repair of satellites and other reconnaissance vehicles.

Although this will probably not continue forever, parts for satellites are almost all made on Earth. But space, of course, offers the best opportunity for testing them. And this takes place in and around the Space Station. Some parts are not only subjected to weightlessness, but also to vacuum and to radiation. These components are therefore installed on the station's exterior, for instance on a type of pallet on which all kinds of experimental apparatus have been placed.

Assembly of separately delivered components can also be done in the Space Station. And something which used to be done by space shuttle or by classic rockets is now taking place on a Space Station: launching satellites. If they are to be located fairly close to the station, then they can be transported by an OMV, an Orbiting Manoeuvring Vehicle, a small compact rocket with relatively little power. If a much higher orbit is to be reached than that of the Space Station, or one in a totally different plane, or even flights to the Moon or the planets, then an OTV will be used — an Orbital Transfer Vehicle. This is a sturdy rocket which places the satellite on the correct course and at the exact speed, then disconnects itself from the satellite, to return again to the Space Station. The rocket is then prepared again for its next

mission. The OMV and the OTV are both unmanned and are controlled from the Space Station. In a special 'garage' these rockets will receive their 'ten million kilometres service' (ten thousand is usually reached within a quarter of an hour!). But satellites can also be recovered by these kinds of rocket. Satellites which become unserviceable due to a minor fault can be retrieved by an OTV, taken to the station, repaired and launched again. Many of the things are done in almost a routine fashion,

and much more cheaply, in the Space Station. The 'garage' or 'hangar' of the Space Station has a fold-away roof, to facilitate the entry and departure of the rockets.

As shown on the large drawing in this book, the Space Station is equipped with cranes and manipulators for picking up and handling satellites and heavy loads. The cranes are also used when building new platforms and extensions to the Space Station. Your accommodation in space is

The OMV (Orbital Manoeuvring Vehicle), a small spacetug which can carry loads to an altitude of 1,850 miles (3,000 km). Controlled from a Space Station or a ground station, the OMV can locate satellites, dock with them, and take the satellites from one track to another, or to and from a Space Station – all within the limits given above. For operations over greater distances and for heavier masses, OTVs are used.

growing all the time, and this is easily achieved by using the module construction system. In the vicinity of the stations, platforms are added regularly, for very special production activities or research.

Right: The end of the manipulator or robot-arm, used to pick up satellites or for launching a fairly light satellite. There is an arm like this to the space shuttle and also to the Space Station. The end of the manipulator consists of a hollow sleeve which slides over a pin on the satellite, so gripping the pin. A small television camera on the manipulator arm provides colour TV pictures to the astronaut operating the arm from the shuttle or Space Station. For manipulating larger and heavier loads, the Space Station also has a real crane.

Top, right: A satellite being launched with the aid of the manipulator. On the left is Florida.

Below, right: A satellite, after retrieval and maintenance, has been given a new lease of life after launching by the manipulator.

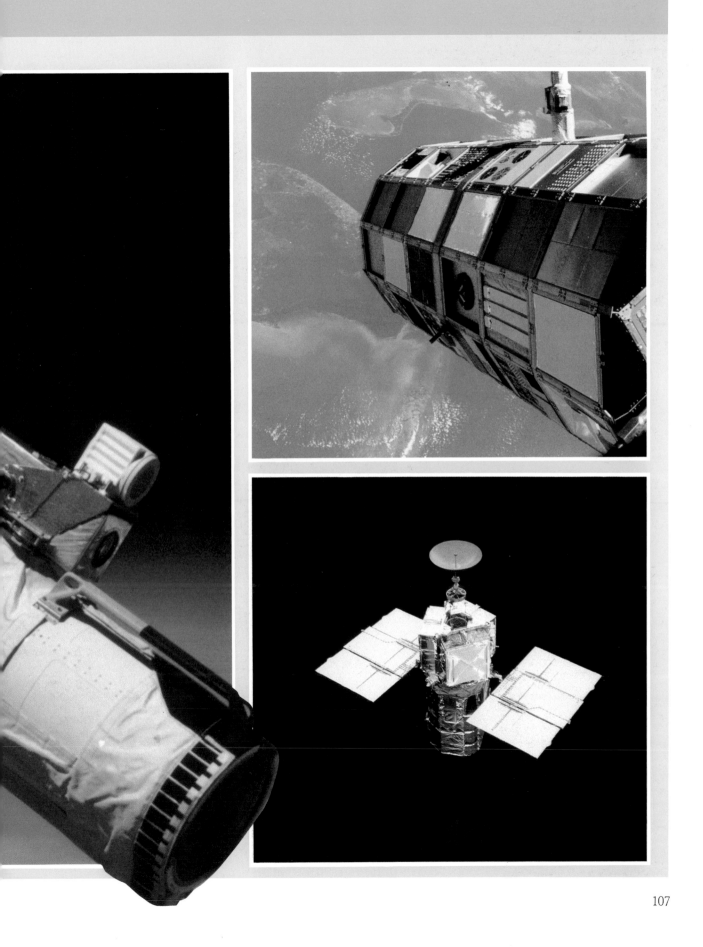

6 GARDENING IN SPACE

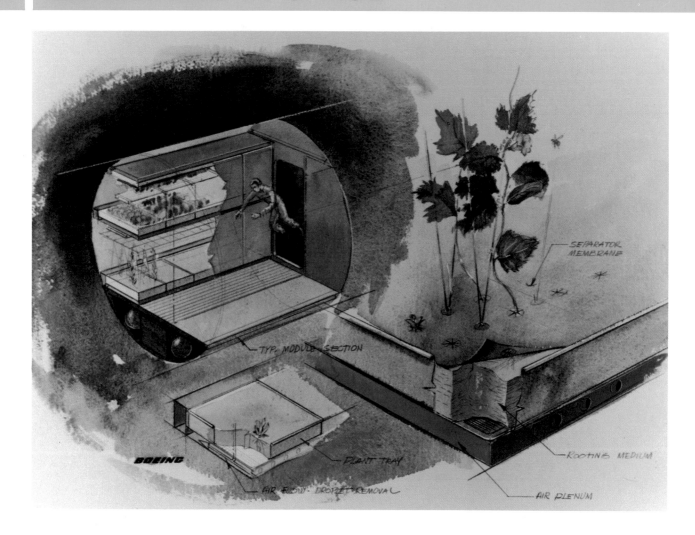

Do you remember when astronauts had to eat primarily out of tubes? Tubes with juice, tubes with mash and tubes with porridge? Yes, it was like that in the first period of space travel. Sometimes the astronauts had a ready-made meal, but that was all. Later on, real sandwiches appeared, wrapped in foil, and also steaks, which could be heated. And these are still there, as recounted in the chapter on food in space. But in the Space Station you can grow a major part of your daily fare yourself in the space garden which is housed in a special vegetation module. Almost all vegetables and many other plants are cultivated here: lettuce, carrots, cabbage, rice: you name it. And the

waste is of course recycled to become manure. Once more, weightlessness necessitates a very special approach to gardening in space, because plants do not want to grow if gravity is absent. Roots, for instance, want to grow downwards, towards gravity, stems and leaves towards the light. If there is no gravity, the roots tend to grow in all directions. So it is necessary to separate the stem and the leaves from the roots, as otherwise the plant grows into a tangled knot. The separation is achieved by using a plastic membrane: the stems and leaves are on top, a root medium and the roots themselves are below.

By this method a restricted number

of plants can be cultivated in weightless conditions. Decorative plants, if handled this way, could brighten up your domicile. In the artificial surrounds of the Space Station these look particularly attractive. But most plants require real gravity, and fortunately the solution is simple. The 'greenhouses' in which are cultivated wheat, radishes, potatoes and curly kale, are shaped like large laundry drums. These rotate slowly, so creating 'gravity' by means of centrifugal force. As the result of the rotation, the plants are pressed outwards, against the inside of the drum. The roots grow outside through the many thousands of holes in the side of the

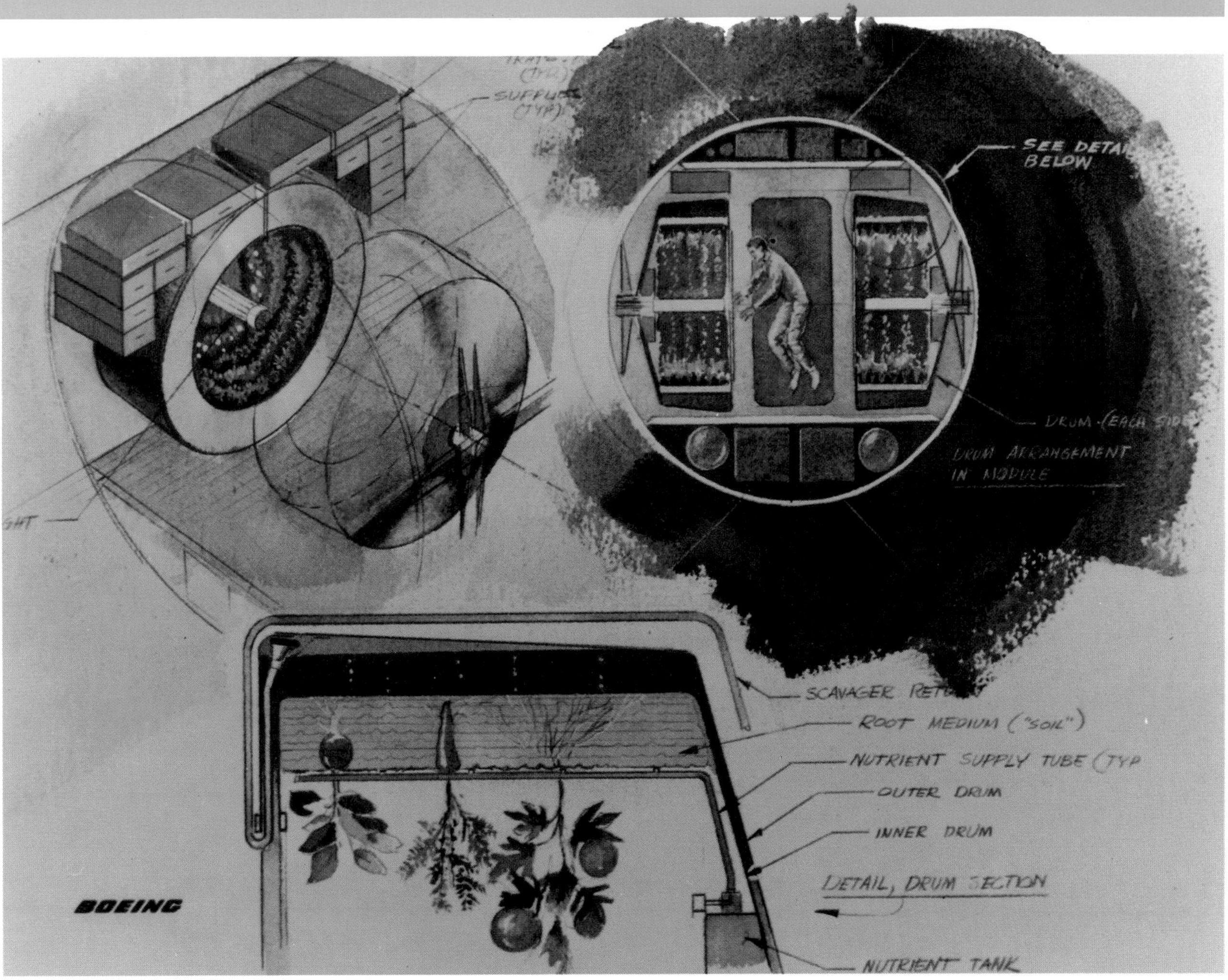
SEE DETA
BELOW
DRUM ARRANGEMENT
IN MODULE
ROOT MEDIUM ("SOIL")
NUTRIENT SUPPLY TUBE
OUTER DRUM
INNER DRUM
DETAIL, DRUM SECTION
NUTRIENT TANK
BOEING

drum. Sometimes a root medium is applied to the outside or — for certain plants — it is sufficient to spray the roots continuously with a plant food solution, in which case no root medium is required at all. The obviously essential light is supplied by neon tubes installed lengthwise in the rotating drum.

This system is not only attractive to you as a space dweller, but is also very cost-effective. In the Space Station, where about 97 per cent of all food is grown, the saving over fifteen years is approximately £50,000,000. Naturally, all human and vegetable waste from the space station are recycled and processed to serve as manure for the plants. Furthermore, the plants consume the carbon dioxide breathed out by humans, and they return oxygen in its place just as on Earth, only on a smaller scale.

Opposite page: How to grow plants in zero gravity: in plant trays: the roots are separated from the stems and the leaves by a membrane. Above the tray are lamps, below the membrane is a rooting medium. Air and water are supplied and drained via the air plenum at the bottom of the tray.

Top: A slightly different version of a garden module for which 'half' rotating hothouses are used on both sides of a central passage.
Underneath: Detail of a rotating hothouse. The plants grow through small holes in a rooting medium.

Below: A tank with liquid nutrient for the rooting medium.

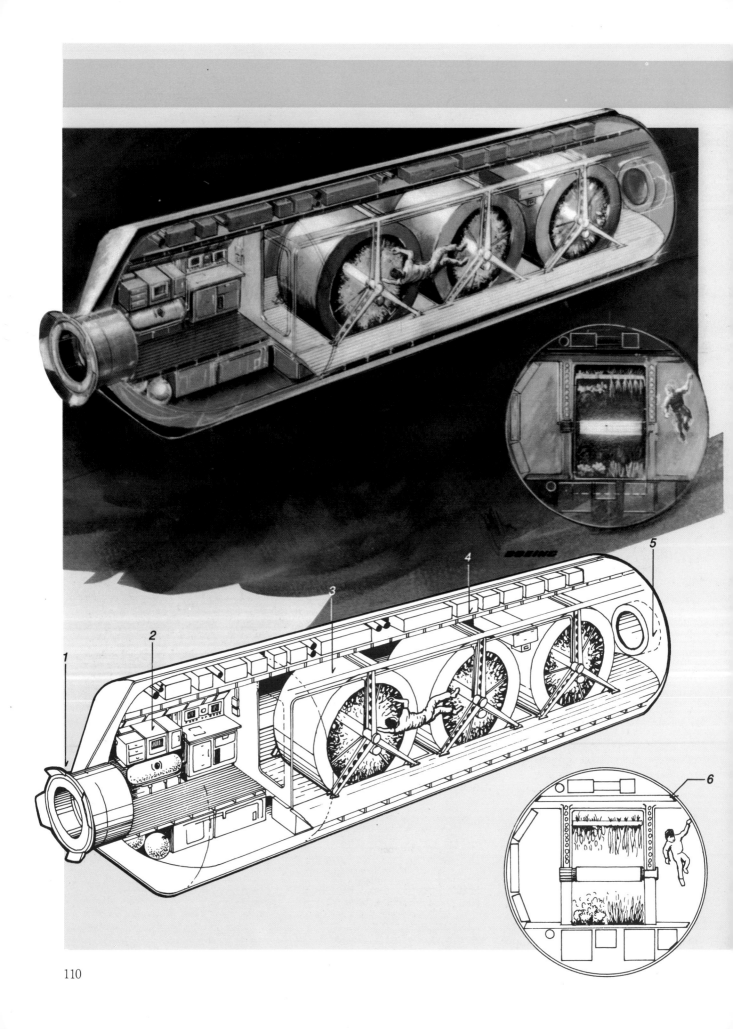

Opposite page: A garden module, using artificial gravity. On the left the cut-away view of the module with three rotating hothouses, on the right a cross-sectional drawing.

Key to the numbers:
1. Connecting tunnel to the Space Station
2. Apparatus for processing harvested vegetables and for recycling refuse.
3. One of the three rotating drums
4. Installation for regulating the environmental conditions
5. Second connecting tunnel for the supply of raw materials and other items.
6. Cross-section view of the module with rotating hothouse.

Lettuce being harvested!

The first attempt to live in space for a longer period ended in disaster. On 29 June 1971 the cosmonauts Georgi Dobrovolski, Vladimir Volkov and Viktor Patsayev died while returning from their 23-day stay in the first space station in the world, Salyut-1. Somehow the air escaped from the cabin of the Soyuz space vehicle, in which the men travelled without space-suits. They died within 7 seconds, and the computer-controlled Soyuz cabin returned only their bodies to Earth.

Their stay in the 100 cubic metre living space of their cosmic caravan (weight 19 tonnes) had been very successful. But the tragedy which occurred at the end resulted in the Soyuz, originally planned for three persons, being used for only two persons for a long time. The cosmonauts then had to wear space-suits during the outward and return journeys, and Soyuz in its original form was just not designed for this purpose. This necessitated the installation of a large ventilation system for the space-suits, which was done at the expense of the third seat. As late as 1979 the modernized Soyuz-T version was introduced, which brought back the three-seat configuration. Room was made by the miniaturization of part of the installation. Due to the delay incurred on the total project, the first 'space-house', Salyut-1, could not be crewed again. After firing its retro-rockets, the vehicle reached the atmosphere on 11 November 1971 and burned up, which also happened to subsequent Salyuts at the end of their useful lives. The Salyut-2 was a failure, yet the Salyut-3 appeared to be extremely useful as a space dwelling.

But this time very few details were released about the space dwelling, which gave rise to the suspicion that it was a military space station, equipped with a gigantic camera for photographing the surface of Earth in great detail. The Salyut was indeed manned by a completely military crew: cosmonauts Pavel Popovitch and Yuri Artyuchin.

Salyut-4 had three independently rotatable solar panels.

The very first Space Station: Salyut-1, launched as early as 1971.

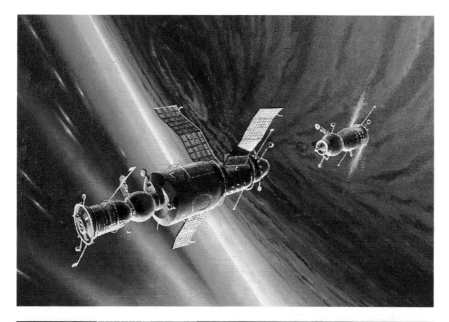

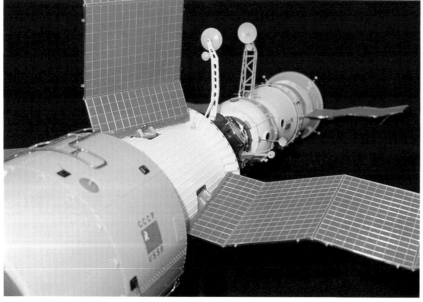

It was quite different with Salyut-6, launched on 29 September 1977. On 9 October Soyuz-25 was launched with cosmonauts Kovalyonok and Ryumin on board was quite different. They did not succeed in linking their space vehicle to the new station. They did, however, make a lasting impression upon the 'front door' of the station which was why this docking port on the Salyut-6 was not used for the next flight. This was possible because this Salyut, unlike its predecessors, offered two docking facilities: one at the front and one at the rear. On 10 December, therefore cosmonauts Yuri Romanenko and Georgi Gretchko floated in their Soyuz-26 towards the second port at the rear of the station. During a prolonged space walk Gretchko inspected the 'front door' thoroughly, and it appeared to be still in excellent shape, so the next expedition was able to dock at that end again.

On 10 January 1978 the Soyuz-27, with cosmonauts Vladimir Dzyanibyekov and Oleg Makarov on board, paid their colleagues a five-day visit. For their return they used the Soyuz-26 of Romanenko and Gretchko, and left their own 'fresh' space vehicle behind for their colleagues. This left the rear entrance free for the arrival of a new twig on the Soviet space tree: the 'Progress', an unmanned space supply vehicle. The Progress — in fact a converted Soyuz — carried about 2,300 kg in supplies and fuel. This fuel had to be pumped via special connections at the rear of the Salyut. After discharging its load, the Progress was jammed full with waste. Finally, this useful vessel was disconnected on 6 February, to burn up in the Earth's atmosphere. The Progress-1 was the first of a long series of freight vehicles used to provide Salyuts with the required supplies at regular intervals and also

Above: Salyut (foreground) and Soyuz, docked together.

Top: A Soyuz space vehicle with crew, approaching a Salyut Space Station. This impression was painted by the Russian artist Andrei Sokolov.

Salyuts-4 to -6

The Soviets were slightly more candid about the Salyut-4, a space station which differed from its predecessors in a number of ways. The Salyut-4 did not have four fixed solar-panels, but three large rotatable panels, allowing the station to move around its centre of gravity without the solar-panels losing touch with their source of energy. The first Salyuts always had to fly with their solar-panels aimed at the Sun. The Salyut-5 was again a military version of the Salyut-4, and very little was seen or heard about it.

to remove all that was unwanted.

The Progress greatly enhanced the possibilities of the Soyuz-Salyut system. Firstly, people could remain on board much longer, because fresh food etc. could be 'mailed' to them. Secondly, large quantities of waste (in a Salyut this amounted to about 20 to 30 kg per day) accumulated indeed during a flight lasting several months and required disposal, even though the Salyut had its own small waste disposal units. The programme became increasingly interesting with the appearance of Progress, because — not least for the cosmonauts themselves — new garments and materials for additional research and experiments could be forwarded to them. Not to be lost sight of was the fact that the life of the Salyuts could be extended considerably by launching Progress space vehicles. For not only did the Salyut require fuel for correction of its orbit and adjusting its attitude with respect to the Earth, but also things like nitrogen and oxygen to maintain the proper atmosphere. The space vehicle was even used to raise the orbit of the Salyut after the spacecraft threatened to get too low due to the small amount of friction at that altitude (about 186 miles, or 300 km) in the very rare atmosphere.

Thanks to the Soyuz-Salyut-Progress system the Soviets were the first eventually to settle in space. This was much in evidence during the four years of the Salyut-6 programme, which was completed in 1981. This space domicile housed in total 43 cosmonauts. On no less than 34 occasions they docked spacecraft at the station. Twelve of these were Progress space vessels, delivering fresh supplies for the crew. Of the manned space flights nine carried an international crew, a Soviet and a cosmonaut from a friendly

An unmanned freightship ('Progress') is docked with the Salyut.

Above: A Russian and a Rumanian cosmonaut during training in a mock-up of Salyut on Earth.

Below: A view of the Salyut space-dwelling. At the back is the panel through which the cosmonauts can enter the transfer chamber. At the other end of the chamber is the Soyuz space vehicle. The backrest of one chair has been removed to show the instrument panel.

socialist country. In sequence, the guest astronauts came from Czechoslovakia, Poland, the German Democratic Republic, Bulgaria, Hungary, Vietnam, Cuba, Mongolia and Rumania.

Salyut-7

The Salyut space dwellings were constantly improved because of the extensive experience gained by the cosmonauts. On 19 April 1982 the Salyut-7 was launched, externally almost identical with the Salyut-6, but internally a large number of modifications had been carried out. The station was used actively for 34 months.

Three times it had a crew who stayed for long periods: 211, 150 and 237 days. Four times, visits were made by three cosmonauts staying

for a week, among them a cosmonaut from France and one from India. And during the Salyut-7 flight a woman took a space walk for the first time — Svetlana Savitskaya, who was also the second woman in space (the first was the Soviet cosmonaut Valentina Tereshkova). The Salyut-7 was the last spacecraft of the second generation (the first generation were stations with only one docking port). After this, the Soviets started to construct the first station of the third generation, a space station constructed of separately launched modules, like the present space stations. The first modular station, launched in 1986, was MIR ("Peace").

The Salyut offered its occupants a quite splendid home, even though it cannot have been a constant delight to live in such a limited space for half a year. However, the space dwelling did have all the modern comforts: a refrigerator, a television monitor (so those on the ground could see the cosmonauts, but the cosmonauts

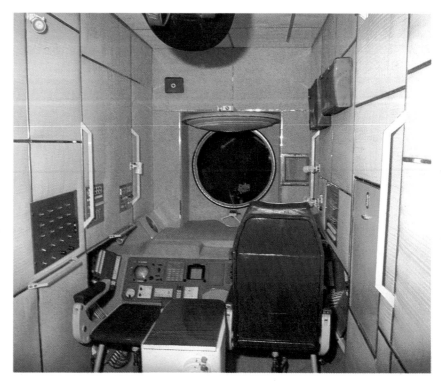

could also now see their families), comfortable beds (against the wall or the ceiling, everything was weightless anyhow), radio, telex, a shower (collapsible), a running track for the daily exercises, a home-trainer (on the ceiling) and a real toilet with a rustic wooden seat. There was also a collection of books and sound and video tapes on board, because the subject of leisure was given a lot of attention for such a long trip. Also, a relatively long time was ·spent on physical exercise: the cosmonauts trained intensively for at least two hours each day. This was very necessary because, as we know, the human body is likely to take it easy in weightless conditions. Heart, muscles and the skeleton are only lightly taxed. The results: a weaker heart-muscle, fewer red blood corpuscles, weaker and thinner muscles and a decreasing percentage of calcium in the bones, which therefore become more fragile. Hence those two hours a day of slavery on the home-trainer, the treadmill, expanders, you name it. For many hours the cosmonauts also wore a so-called 'penguin-suit', a suit with built-in springs which opposed the muscles and so exercised them.

On Saturdays and Sundays the cosmonauts were off duty. They could then read, or listen to their favourite music, watch television or talk to their relatives and friends on Earth. In this way they had very close contact with Mother Earth, something highly necessary in order to 'hold out' for such a long time in such a restricted environment.

One of the beds in the Salyut, against the ceiling. In the centre is a waste-sluice, and on the left the collapsible shower bath.

Above, the home-trainer on the ceiling of the Salyut.

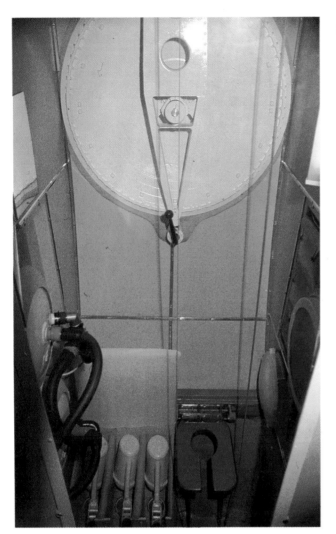

The toilet in the Salyut, complete with rustic wooden seat. Each cosmonaut has his own little pot (to the left of the toilet). Above the toilet is the panel giving access to the rear entrance of the Salyut, through which everything on board a Progress freight ship is passed into the Salyut.

Cosmonaut Vladimir Dzyanibyekov floating inside the Salyut-6.

Cosmonauts Valery Ryumin (left) and Leonid Popov during training. Ryumin remained in space for over a year, spread over two space flights on board Salyut-6.

Results

What were the results of these Salyut flights? The cosmonauts studied the Universe, and, so to speak, took an inventory of Earth, of both the land and the sea. In an era of increasing concern at the world-wide lack of raw materials and energy this is not unimportant.

Travelling through space seemed the only way to compile such a world-encompassing inventory at a reasonable cost and within a reasonable length of time. This inventory was made — and it is still going on — in the main by means of photography. An instrument much used in the Salyut was a six-lens camera, produced by Zeiss Jena in the GDR. At each click of the shutters a fairly large portion of Earth was photographed in six different colours. And each colour was sensitive to a particular property of an area down below. Thus it was possible to depict all foliage — in fact all living plants — in dark red.

Speaking about the photographic possibilities of manned and unmanned space travel, the Soviet Professor Andrei Petrovitch Kapitsa stated: 'The most interesting part of cosmic photography is that one can see much larger parts of the Earth. If you make a mosaic of **aerial** photographs, then you won't see as much, at least of large-scale objects. The pieces forming the mosaic are each very different in shadow and colour. The edges do not fit together properly. Much information is lost in this way. We discovered that if we placed such a mosaic next to a space photograph we saw much more on the latter.'

With the aid of a Cosmos module (foreground), the living space of the Salyut-7 (background) was temporarily doubled. Similar modules play an important role during the construction of Space Stations from separately launched blocks.

After a record-breaking flight of half a year, cosmonauts Ryumin (left) and Lyakhov have a little difficulty in getting used to Earth's gravitational force. On the right is the Soyuz space-landing capsule.

At first, it seems odd that using space photographs you can discover coal and oil. Kapitsa stated: 'There are methods to do so, even if it sounds crazy. In the first place I want to mention the composition of the surface of the land as we see it from space. Geological maps, made using space photographs, show the geologists where to look for minerals. They do not have to search over a large area to try to find fault zones. Moreover, they can look for pointers: certain plants are often indicative of various types of soil and minerals. The colours of the plants depend on the type of soil. By analysing the colours of the plants one can find minerals. Via the vegetation you can actually see what is underneath them. You can also use microwaves to determine the temperature and the humidity of the surface of the Earth. So, in a certain way, you can indeed look beneath the surface of the Earth.'

From Soviet publications which appeared during Salyut era the significance of the study of Earth and

its environment became more and more evident. The observations carried out on board the Salyuts were used by hundreds of organizations in the USSR. Through this space detective work, for instance, oil was located in many places in the Soviet Union. Space photography was also very important to agriculture, permitting an inventory of the harvest and the discovery of diseases in crops

long before they became noticeable on Earth.

Back in the sixties it was suggested that it should be possible to detect shoals of fish from an orbit around Earth. In 1978 Vladimir Kovalyonok carried out the first observations to detect fish from on board the Salyut-6. Of course, the fish cannot actually be seen from space, but large concentrations of food for the fish can be detected in the water, while the temperature of the water can also have some relation to the presence or absence of fish. In a great number of cases these types of observations have led to definite success. The cosmonauts Lyakhov and Ryumin discovered large concentrations of plankton during their space marathon on board Salyut-7 in 1979, and communicated the exact positions back to Earth. Soviet fishermen in the Atlantic Ocean then discovered a gigantic shoal of mackerel. In another instance fisherman in the Pacific Ocean caught a large quantity of squid after a tip from space. Currently, such observations are passed on quickly and routinely.

Fishery and shipping started to benefit in the seventies in yet another way. Professor Kapitsa said: 'Meteorological satellites provided us with information to set out optimal routes for the ships. We can measure the height of the waves through them and, for example, look at the ice cover on the northern passage. Earlier, we needed hundreds of flights by aircraft to record all the changes. Now the "Meteor-system" provides us with all necessary information, and that in a very economical way. Our meteorologists can now offer the captains optimal routes, avoiding cyclones and high seas whenever possible. Through this the average speed of our ships can be kept as high as possible. Many ships already receive their information direct from the satellites.'

The extensive forests in the Soviet Union are continuously threatened by fire, particularly in the hot season. These were also reported by the Salyut cosmonauts at an early stage to the Flight Control Centre. The same applies to hurricanes. A sand storm which started in Central Africa was tracked by Lyakhov and Ryumin

across the Atlantic up to the American east coast. Timely warnings of such catastrophes could save many lives and much material.

The many thousands of photographs taken from the Salyut also contributed to better maps. Of course, aerial photography has led to enormously improved mapping, but only from space can we review Earth in its totality. It is not surprising that photographs made by the cosmonauts were used to up-date maps already in existence, and to modify coastlines which are always changing. But more specialized maps were also produced. The cosmonauts provided the basic material for an all-embracing snow and ice chart of the world. If the Sun — or an increase in the average percentage of carbon dioxide in the air — should cause the average temperature on Earth to increase by one degree, then extensive areas would be flooded by the melt-water. Over a longer period this danger is not just imaginary — particularly because of the large quantities of carbon dioxide

produced by industry. That is why it is of the utmost importance to follow closely the developments in the masses of snow and ice. Using Salyut's photographs a tectonic map of the Soviet Union was produced, clearly indicating the fault lines, so that better predictions are possible of the chances of an earthquake. Also based on the Salyut's photographs, an inventory of the available timber in the Soviet Union was produced. If all these observations had been made by aircraft it would have cost an unbelievable amount of time and money. The super-camera installed in the Salyut 'snapped' large areas of Earth (some hundreds of square kilometres) in as little time as it takes to blink an eye.

Apart from this camera, the Soviets also used an enormous telescope with which parts of the Universe and Earth could be photographed. The reflector of this telescope had a diameter of no less than 5 ft (1.5m). This instrument was not only used to photograph stars, but also tropical cyclones were tracked along their routes.

Then, of course, there was the research on the Universe itself. A persistent myth is that space is empty. Well, there is no truth in that — space is filled with electrically charged particles, most of them originating from the Sun, but also from deep space between the stars. These particles have quite a considerable influence upon Earth's atmosphere. Among other things, they cause the polar-light phenomenon. But the so-called solar wind is also responsible for interference to radio communications and strong variations in the magnetic field of Earth. More knowledge about this particle radiation would be welcome.

From the Salyut — way above the interfering atmosphere — cosmonauts also studied the Sun and other stars. This was done by means of sensitive measuring instruments, cameras and telescopes. New experiments were also conducted in the sphere of producing new materials in space. In their 'Splav' furnace the cosmonauts produced more than a hundred different alloys, metal-mixtures whose manufacture had been thought impossible. Due to the absence of gravity on board the Salyut it was indeed possible to produce them. The cosmonauts also made some unique and very pure crystals, which can be used for the production of highly sophisticated electronic apparatus. They also produced glass lenses for high-grade optical instruments which, after having cooled off, were ready for immediate use and did not require any subsequent grinding or polishing. In this way Salyut's occupants laid the foundation for later — large-scale — industrial activity in orbit around Earth.

During these very long-lasting flights in the Salyuts, human life itself withstood an important test. It was proved that humans could live at least half a year in space without detrimental results. Thanks to the Salyut flights we learned that Man is spaceworthy; that Man could make a journey to Mars, because a one-way trip would take half a year. But, most important of all, we learned that Man can do something useful in space.

In the Salyut's transfer compartment: Svetlana
Savitskaya and Vladimir Dzyanibyekov.

Svetlana Savitskaya at work
outside the Salyut-7.

Far left: In the cosmonaut
training centre; Svetlana Savitskaya
together with Igor Volk (left)
and Vladimir Dzyanibyekov.

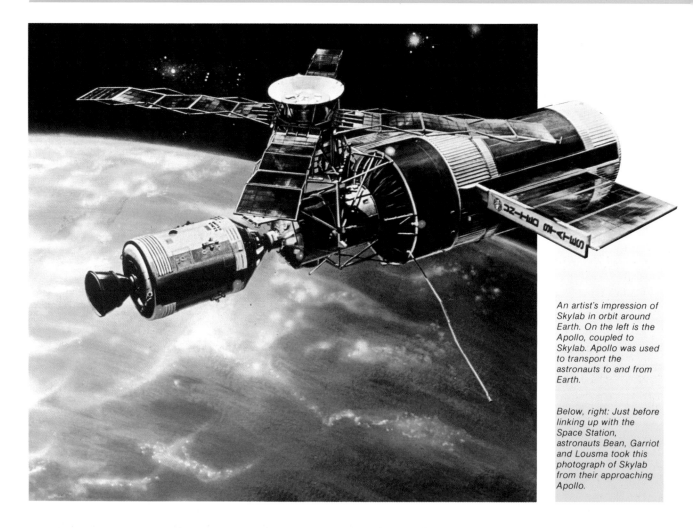

An artist's impression of Skylab in orbit around Earth. On the left is the Apollo, coupled to Skylab. Apollo was used to transport the astronauts to and from Earth.

Below, right: Just before linking up with the Space Station, astronauts Bean, Garriot and Lousma took this photograph of Skylab from their approaching Apollo.

If the Salyut could be classed as a 'space bungalow', the American Skylab was a true 'space house', which offered no less than 347 cubic metres of living space. This space station weighed 90 tonnes. Unfortunately, there was only one Skylab due to the curtailing of NASA's budget. It was manned three times, in 1973 and 1974, by astronaut threesomes. The first crew, consisting of Charles Conrad, Joseph Kerwin and Paul Weitz, lived 28 days in the enormous (for its time) space station; the second trio of Alan Bean, Owen Garriot and Jack Lousma stayed for 56 days; and the third, Gerald Carr, William Pogue and Ed Gibson, lived for 84 days in Skylab.

This record was subsequently beaten by the Soviets in their Salyuts. Unfortunately, Skylab could not be manned again due to the tight strings on NASA's expenditure, although the station stayed in excellent condition for a long period afterwards. On 11 July 1979 it was lost in Earth's atmosphere, because the space shuttle could not be made available in time to give the colossus a little push upwards. So ended a particularly successful project, the first manned American space programme that was significantly useful to our earthly existence.

Rescue plan

However, immediately after the launch of Skylab on 14 May 1973, the project looked as though it was going to fail. During the launch, on top of the large Saturn-5, a large piece of the outer insulation was ripped off, together with an extending wing with solar cells which was to have supplied electricity.

A second, identical wing did not unfold when the spacecraft had reached its orbit. The result was that only a few small solar panels could supply electrical power, while the temperature inside the station rose to dangerous heights due to the missing insulation. The launch of the first three occupants, Charles Conrad (who had already been to the Moon with Apollo-12), the physician Joseph Kerwin and newcomer Paul Weitz, had to be postponed as a result. They should have been launched one day after Skylab in one of the Apollo capsules, left over after the premature ending of the Moon programme. Now they had to wait on the ground for another ten days, while feverish activity went on to develop a rescue plan for the crippled spacecraft. This boiled down to the astronauts themselves having to make their space dwelling habitable again. By a miracle this rescue plan was 100 per cent successful. Conrad and his crew deployed a parasol, sprayed with gold, over Skylab, so that the temperature went down. And they were able to free the locked wing with its solar cells, using a fair-sized pair of steel cutters during a space walk.

Construction

Including the Apollo coupled to it, Skylab consisted of five major components:

— The Apollo, taking care of delivery and return of the crews

— The docking section to which, in an emergency, two Apollos could connect (which did not, however, occur)

— An airlock, as a 'porch' to the space station

— A working and domestic space of two storeys. Along its exterior two large solar panels were installed, one of which was lost during launch.

— A Sun observatory, with four smaller solar panels.

Skylab was mainly constructed from components which had been developed for the Apollo and Gemini projects. The working space was

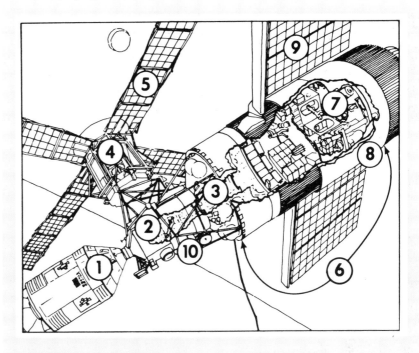

The major components of the Skylab Space Station

1. Apollo
2. Docking adaptor
3. Air lock
4. Solar telescope
5. Solar panel, Solar telescope
6. Habitat-working space on two levels
7. Habitat compartment
8. Third stage of Saturn rocket
9. Solar panel

actually the third stage of the Saturn-5, in this instance not active! In the context of the Moon flights this stage was used to push a combination of mother-craft and Moonlander in the direction of the Moon.

Skylab was only required to achieve a relatively low orbit at just over 250 miles (400 km) altitude, so the first two stages of the gigantic Saturn were alone quite capable of placing the station in the required orbit

The comfortably equipped working and habitat modules in Skylab were accommodated in what used to be the tank for liquid hydrogen of the third stage. This stage also contained a smaller tank for liquid oxygen, which was only used as a king-sized garbage can by the astronauts.

The working and domestic space had two storeys. In the top was the equipment for various experiments, and food and other stores. In this part, which was not furnished, there was so much room that the astronauts could perform all kinds of capers and could also test a forerunner of the 'space scooter', the MMU.

The lower deck of the habitat and working spaces, if you could describe it thus in weightless conditions, housed the crew quarters and the installations required for mainly medical experiments. The three astronauts each had a small bedroom. There was also a dining room, where hot meals of normal composition could for the first time be prepared during a space flight.

The lower floor of Skylab was also equipped with an installation for the processing of all kinds of waste. An unheard of luxury for the astronauts was a real toilet. Until the Skylab era they had had to make do with plastic bags. And once a week the men could now take a real shower in a cylindrical, bellow-shaped bag, which

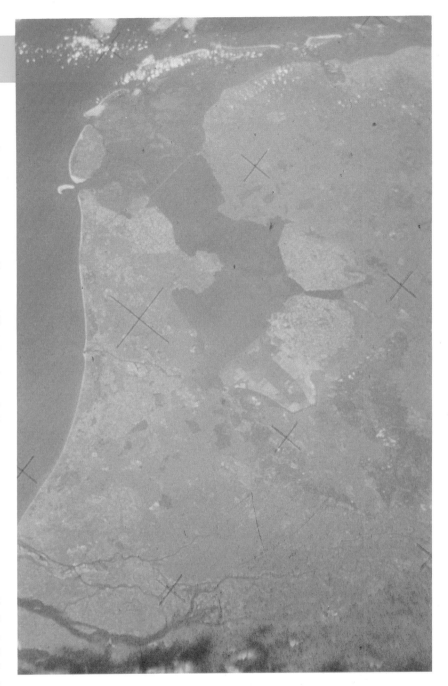

The Netherlands, seen from Skylab.

Opposite page: During a walk in space an astronaut frees the space station's jammed solar panels, using a large pair of cutters.

could be sealed completely so that no weightless water droplets escaped. At the request of the astronauts, who liked to look at the views every now and then, another window was installed in the wall of the dining room, shortly before the launch.

Above the working space in Skylab (in the launch position), an airlock had been installed. This was divided into three sections separated by ports. The primary function of the airlock was to isolate the docking section from the working space, so that

separate 'cells' were created. This was to obviate the possibility of one of Skylab's compartments springing a leak and the air in it escaping. The centre airlock section contained an 'outer' port in the shape of a panel left over from the Gemini programme.

It was via this port that an astronaut had to go outside to exchange film cassettes in the big solar telescope which was installed on the side of Skylab.

The docking section also held the control panels for the photographic equipment pointed at Earth, and the large control panel for the solar telescope — in fact a complete battery of instruments for research on the Sun and for photography — and it weighed no less than 10 tonnes. During the launch this installation sat on top of the docking section, to be unfolded outwards into space. The four solar panels which powered the installation were also unfolded at that time.

Research

The research carried out in Skylab was more or less the same as that conducted in the Soviet Salyut space stations. It can be divided into three main categories:

— MEDICAL EXPERIMENTS Here the chief object was to determine if Man could sustain a stay in space of 28 to 84 days, and if the subsequent return to Earth would be possible without problems. Human beings indeed appeared not to suffer from insurmountable problems after a lengthy stay in space.

— RESEARCH INTO THE EARTH AND ITS NATURAL RESOURCES For this research Skylab had been equipped with a number of specialized detectors and cameras which could register both visible light rays and rays at different wavelengths (particularly infra-red rays). With the aid of these instrument packages, the so-called Earth Resources Experimental Package (EREP) data was collected for agriculture, geology, geography, hydrological control on Earth and research into the environment in general.

— SUN RESEARCH AND OTHER SCIENTIFIC EXPERIMENTS The Sun has great influence upon the Earth's atmosphere and is our main source of energy. The astronauts paid a great deal of attention to ultraviolet rays and X-rays from the Sun, which, due to the protective layer provided by the atmosphere, can barely be detected from the surface of the Earth. The stars were studied too. Particularly interesting was the investigation of certain manufacturing processes (for instance production of alloys and crystals) at zero gravity.

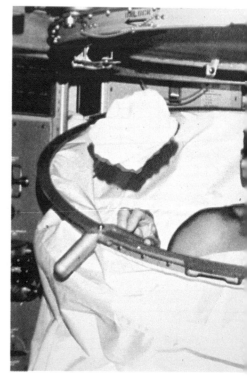

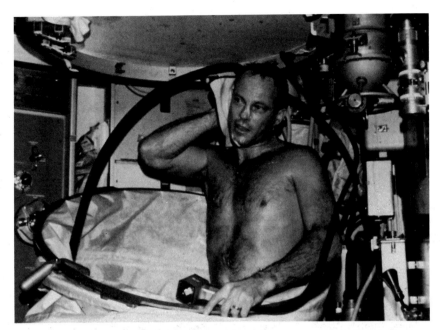

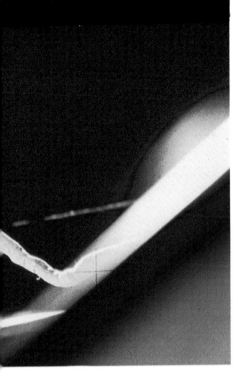

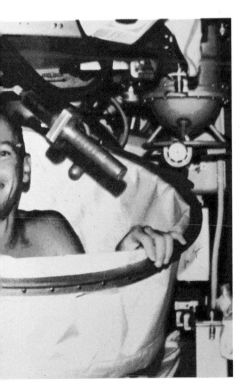

Past: Spacelab

With Spacelab, Europeans were for the first time given the opportunity to live and work in space. During the seventies, Europe constructed a small space laboratory that could be launched in the cargo bay of the space shuttle and which remained there during the flight, which was to last for a maximum of ten days. Two to four persons could work in this high-level laboratory. For the supply of gas, water and electricity Spacelab had to depend totally on the space shuttle. In any case, Spacelab only served as working space for the astronauts: their time off was spent in the crew compartment of the shuttle. Through the open cargo bay door of the shuttle the crew (partly made up of Americans) could look at Earth and the Universe, and study them.

The people living on board Spacelab were not the classic astronauts, the pilots mainly coming from the military forces. No, these were scientists. And thus the first clear step was taken in the direction of the democratization of the space business.

Box of bricks

The Spacelab design was very well planned. It was founded on the principle of a box of bricks: as you wished, you could add or remove a block. The lab was formed by two main components: a pressure cabin in which people could work in their shirt-sleeves, and a kind of open box, comprising pallets on which instruments were placed.

People cannot do without air, but this gaseous mixture is not very kind to telescopes, cameras and sensitive instruments. All the special instruments were controlled from the inside. Of course, every now and then a flight specialist had to go outside to exchange a cassette or to adjust something. The real scientists, the so-called 'payload specialists', were not allowed outside.

Both the pressure cabin and the open cargo bay could take a maximum of five pallets, but then no room would be left over for the pressure cabin. Therefore the instruments would have to be highly automated, and, where required, controllable from the 'bridge' of the space shuttle. The pressure cabin itself could be made up of one block, or two, or three, as required.

So with Spacelab you could have it whichever way you wanted it; the more so as all instruments in the pressure cabin had been mounted in racks which could easily be exchanged. Actually, the instruments could literally be wheeled out of Spacelab after its return to Earth — just like that. It was, therefore, very easy to 'tune' the interior of Spacelab completely for a new flight with new experiments. Each Spacelab block was calculated to be used for at least fifty flights. During the launch and the return to Earth, the scientists just sat in the spacecraft's cabin. The few off-duty hours they had, they also spent in that cabin (on board Spacelab they worked for twelve hours a day!).

Top, left: ESA astronauts Dr Ulf Merbold (left) and Dr Wubbo Ockels.

Opposite: The take-off of Spacelab-1 inside the space shuttle Columbia, 28 November 1983.

Inset: The crew of Spacelab-1 en route to the astronaut's bus which takes them to the launch site.

Research

The scientists on Spacelab occupied themselves with four categories of research:

— THE STUDY OF EARTH Via cameras and measuring instruments the Spacelab workers, from their grandstand seats, received important information about Earth which could be applied to such diverse subjects as transport, town and country planning, environment control, agriculture, fishing, navigation and meteorological forecasting. As we saw before, it is also possible, from an orbit around Earth, to gather information on the presence of minerals. The scientists on board Spacelab also tested new instruments, which were to be installed later on, e.g. automatically functioning environment monitoring satellites.

— ASTRONOMY The interaction between the Sun's radiation and the environment on Earth is very important. But stars, comets, planets and constellations can also be observed much better from space than from Earth.

— MEDICAL SCIENCE IN SPACE, AND DRUGS We learned much, thanks to Spacelab, especially concerning the influence of weightlessness on the organism (particularly the effects over short periods). Tests were also carried out with a view to the production of drugs in space.

— INDUSTRIAL PROCESSES The lessons learned about the production of new materials during weightlessness, as achieved in Salyut and Skylab, were further extended in Spacelab.

Flights

The first flight of a Spacelab started on 28 November 1983, when the space shuttle 'Columbia' lifted off with the European Space Laboratory in its cargo bay. The flight, originally intended to last for nine days, was in fact to take ten days.

The commander of this mission was veteran John Young, who was carrying out his sixth flight, and who had also flown the first flight in 1981 with the space shuttle. He was also one of the twelve persons who, during the Apollo programme, had set foot on the Moon. The very young-looking Brewster Shaw acted as co-pilot.

Left: This is how the shuttle (with Spacelab still on board) travelled around Earth.

Owen Garriot (left) and Ulf Merbold, first ESA astronaut in space, at work in Spacelab.

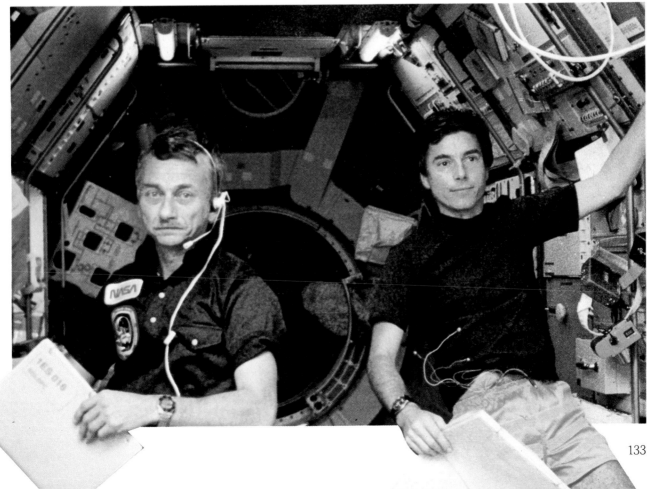

Inside Spacelab the work was carried out by a total of four astronauts, three Americans and one European. The three Americans were Owen Garriot (a Skylab veteran), Bob Parker and Byron Lichtenberg. The European was ESA astronaut Ulf Merbold, originating from Germany and a colleague of the Dutchman Wubbo Ockels, and the Swiss Claude Nicollier. Dr. Wubbo Ockels was Ulf Merbold's primary liaison officer in Houston during the flight.

During this maiden voyage of Spacelab more than seventy experiments were carried out, in seven different fields of research: human science (biological and medicinal), materials, solar research, astronomy, Earth observation, atmospheric physics and plasma physics. The voyage conclusively proved the value of Spacelab as a research institute at high level. Columbia and Spacelab returned to Earth on 8 December 1983, at Edwards Air Force Base in California.

During a subsequent flight, which was largely paid for by West Germany and therefore named the 'D-1 mission', the Dutch astronaut Dr Wubbo Ockels went into space. Together with two German scientists he carried out research.

Spacelab was used, in various configurations, to carry out many more flights. The successful Spacelab design proved to be a solid base for further studies carried out in Europe into the manufacture of building blocks for large Space Stations. These elements, developed within the Columbus project, not only led to European participation in the American Space Station, but also to a European Space Station in orbit around Earth.

Top: Columbia with Spacelab returning from space, 8 December 1983.

Top, right: Wubbo Ockels during flight training in the European space laboratory.

Right: The American Byron Lichtenberg conducting research on materials on board Spacelab.

Far right: The Spacelab crew leaving the shuttle.

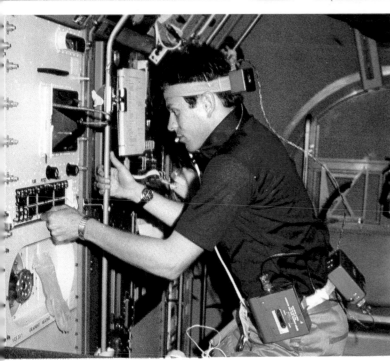

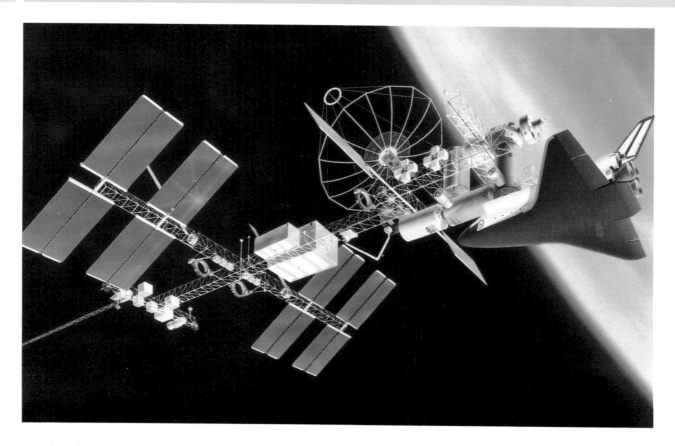

On 12 April 1961, the first human being — Yuri Gagarin — survived away from Earth for just a short time: one and a half hours. For hundreds of millions of years everything on Earth had been bound by the fetters of gravity to the surface of our planet. But now Man had freed himself to live in the adverse conditions of space.

By 1995 hundreds of people have sojourned in space — some even for many months. Man has therefore started to live in space. At first this occurred in rather cramped space environments like Salyut and Skylab: Space Stations of a few hundreds of cubic metres at most. But then appeared modular space dwellings, constructed of separately launched blocks, like Space Stations. It is quite pleasant to live here, even for long periods. And, above all, it is interesting to live away from Earth, free from gravity.

But weightlessness also brings with it the unavoidable difficulties, especially if longer periods are involved. Human beings developed on Earth, and have always been subjected to gravity. And it is evident that we cannot do without weight in the long run. Which is why the space dwellings of the future will differ from the space stations of today: they will have artificial gravity, so that the occupants do not have to whirl around in disorder, however much fun this might be for a while. But, above all, in this way the negative effects of long periods of weightlessness (like the loss of calcium and weakening of the muscles) are prevented.

'Spin dryer'
We have known for a long time how to create that artificial gravity. Long before the launch of the first Sputnik,

space prophets like the Russian Konstantin Tsiolkovsky (1857-1935), the Austrian Herman Oberth (born 1894), and the German Wernher von Braun (1912-1977), had pointed out that the principle of centrifugal force offered the solution to this problem. Future space stations will have the shape of a gigantic wheel, which slowly rotates around its axis. The accommodation for the occupants will be in the 'tyre' of the wheel and, due to the centrifugal force, they will be pressed against the outer wall with their feet. The architects will, of course, make sure that the floor will be in that position. If they move from the 'tyre' via an elevator through a spoke, the occupants will feel themselves getting lighter as they get nearer to the axis. When moving in the opposite direction they will gradually become heavier. The centre part, the axis, could be used for

Left: In 1995, human beings on board Space Station will still have to do without gravity.

Right: The more one thinks of living in space for any length of time, the more necessary it will become to consider the use of artificial gravity.

Below, right: The space shuttle about to link up with a Space Station orbiting Earth. On the right, a free-flying platform used for experiments and production in which the presence of humans could be a disturbing factor.

accommodating laboratories and other rooms where weightlessness is essential. Certain relaxation rooms, like a 'swimming-pool without water', could be part of these.

Rotating space stations with artificial gravity must, by definition, be big, as otherwise the occupants would become dizzy due to the merry-go-round effect. A diameter of one to two kilometres is therefore the minimum. When we think of such large structures the problem of their basic materials becomes important. Because Earth has a lot of gravity and an atmosphere, it requires much energy to launch one kilogramme of basic material from Earth.

Construction in space would therefore be very expensive. But we can do something about that. Not all that far away we have an enormous source of basic materials: the Moon. As a result of the Apollo flights we now know what we can find there:

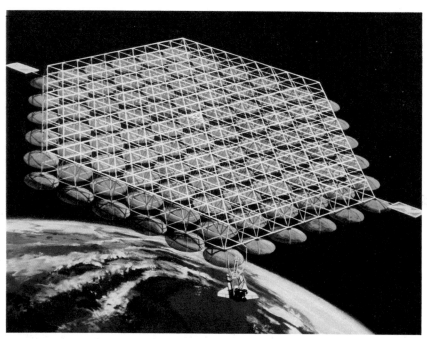

The crew of a space shuttle constructing a gigantic platform for electronic communications, which could cater for communications between continents on a large scale.

aluminium, titanium, glass, and oxygen (trapped in the rocks). So only hydrogen, carbon and nitrogen will have to be supplied from Earth. The Moon has light gravity, one sixth of that on Earth. It is easier therefore to lift basic elements from the Moon.

This can, however, only be done if we have a mature space transport system, which could be as follows. Between Earth and the space stations a regular service is maintained by space shuttles and their development. From these space stations, shuttling rockets could fly to lunar space stations, orbiting the Moon, at about 250,000 miles (400,000 km) from Earth. From there we can go, with special landing craft, to the pockmarked surface of the Moon.

All elements of this system can be used again and again. To begin with, this seemingly adventurous journey will be made by some 150 Moonminers. They will construct installations for scooping up basic elements, and will maintain them. They will supply us with practically all the basic elements required. The materials required from Earth (of which hydrogen is the most important) will only be required in relatively small quantities: we can re-use these repeatedly. Hydrogen will also be used to extract oxygen from Moon-ore, so as to form water. Part of that water can later be split up (by electrolysis), the hydrogen regained, and oxygen then becomes available for breathing. All life in the Moon colony and in our space colonies will take place in an almost ideal recycling process, so that there are hardly any losses, and only little is required in addition.

Once the mining is going full blast we can take the next big step: from space stations to complete villages and cities, floating freely in space.

Above this first village in space is a floating mirror to ensure that sunlight is divided equally over the windows of the space village, which are installed along the inside of the 'rim'. Blinds provide darkness should the inhabitants require it.

Vision

A futuristic vision: a silver cylinder, 20 miles (32 km) long and more than 4 miles (6 km) in diameter, slowly rotates around its axis, hanging in space at about 250,000 miles (400,000 km) from Earth and Moon. Miniature craft flash to and fro like busy bees between this space colony and oddly shaped factory sites in its vicinity. This picture of immense settlements in space seems to attract more and more supporters. Admittedly, it was outlined long ago by enthusiastic space-travel prophets. But in the seventies a large group of specialists, including some from NASA, started to consider the subject seriously for the first time. This took place under the direction of Professor Gerard O'Neill, of Princeton University. The conclusion formed by that esteemed body of technologists, psychologists, economists and sociologists was that it is as unavoidable as it is desirable that we go into space — on a large scale.

In the middle of the next century the first space colony could be a reality, a relatively small settlement with about ten thousand inhabitants. But this is only a modest beginning. In the next era, the group envisages the space around Earth enriched by larger colonies, of which pictures are shown above, for up to one million inhabitants.

For our descendants there will be many more opportunities than for us to 'get shut' of Earth, and even to settle permanently in space.

According to Professor O'Neill and his supporters, technology in space should not be used to create a cold

A view of a space village. There is no need to feel confined, because you can see for more than 800 metres.

world based on metal and plastics, like we meet so often in science fiction. No; familiar, earthly surroundings are to be imitated whenever possible.

In this respect the space regions of the future will differ greatly from today's space stations. Of course, artificial gravity will prevail inside the space cities. But eye-catching landscapes will be constructed which will give the observer the impression that he is still on Earth.

Space village

The construction of the first space region, a modest village for ten thousand inhabitants, will start from a rather simple space station which circles Earth in a wide orbit, and which acts as 'site hut'. It will be manned by about one hundred space construction workers who, of course, have been taken to the station in a passenger version of the space shuttle. People will be able to travel (with about just as much comfort as today's jet passengers) in the cargo bay of the shuttle, albeit that they will be weightless for nearly all the time — which will not be long — that the flight will take. The sign 'fasten seat belts' will be on during the whole journey to avoid excessive confusion!

The lion's share of the building materials will come from the Moon, as we said earlier. From the 'site hut' the construction workers will have mobility using a 'space-tug', a craft operating only in space, and which, therefore, is considerably lighter and simpler than a space shuttle.

The destination is a point 250,000 miles (400,000 km) away from Earth and Moon, where the first village in space will grow. This point, together with these two heavenly bodies, forms an imaginary isoceles triangle in space. A body in this position will always maintain its position relative to Earth and Moon, and this is very important from a regular communications point of view.

Once on the construction site, the construction workers start by building the steel factories, of which the prefabricated elements will have to be transported by the space

The next phase: A large spherical space colony for 50,000 to 100,000 inhabitants.

Below, left: The interior of the spherical space colony. Near the imaginary axis there is little gravity, allowing the use of an 'air bike'.

Below, right: Two large cylindrical colonies which together could accommodate one million inhabitants. The petal-shaped strips are mirrors for the division of sunlight and also serve as blinds for the windows.

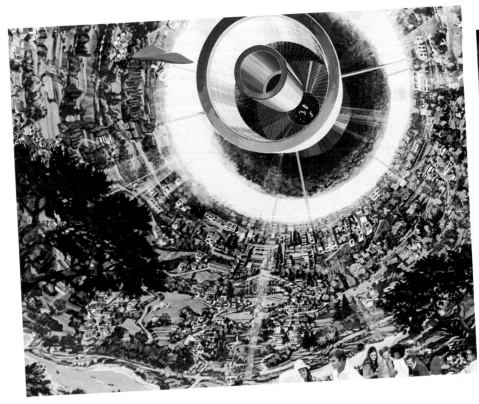

shuttles and the space tug. But how to remove the ore from the Moon? A simple solution was found, for it is very easy to do, due to the low gravity. A kind of enormous 'dredger' will be used, which launches an uninterrupted stream of basic elements in the direction of the space factories on the construction site. Imagine a race-track, several kilometres long, on which are placed the full trays of ore. These are accelerated until they have reached the Moon's escape velocity (about 2.5 kilometres per second). At that moment the open trays are stopped, but their contents continue to sail into space. At the construction site a kind of arrester net slows down the material, which is then fed into an outsize funnel. The space factory then converts the ore into aluminium, titanium, iron, glass, oxygen and so on.

The first space colony will have the shape of a wheel, with a diameter of about two kilometres, rotating once per minute. At this speed the space dwellers experience precisely their own weight. The 'tyre', housing the people, is almost 650 ft (200m) thick. This first cosmic settlement will have an earth-like landscape and the inhabitants will be assured of a view of at least half a mile, in order not to feel claustrophobic. The 'tyre', in which all domestic activity takes place, will be divided into separate sections. To begin with, this is for safety, because there is a — minute — chance of a meteorite strike, causing the air to escape. A division into sections is also essential due to the different applications of the various sections. Some parts will house inns and shopping centres, in which there will be a climate comfortable for humans. But other sections will be for agricultural purposes. It may be that here the blinds in front of the windows will never be closed, to let the sunlight in at all times. The percentage of carbon dioxide will be increased considerably, because plants consume carbon dioxide and release oxygen, which again can be breathed in by humans. And humans exhale carbon dioxide, which is useful to the plants. With an excess of carbon dioxide and light the plants will flourish, and many crops will be possible per year. It goes without saying that we can completely control all aspects of the weather in these space regions.

Space colonies

The first space village will be minute compared with the large space colonies which will be realized in the latter half of the coming century. According to O'Neill, they will be 20 miles (32 km) long and 3.7 miles (6 km) in diameter and, depending on their internal arrangement, offer accommodation for between one hundred thousand and one million people. A gigantic cylinder like that

Like Paradise: A corner of the large space colony. A lovely spot to spend a weekend.

141

has three long landscapes along its inner wall, separated individually by large windows along the full length of the cylinder. Moving blinds ensure that the rhythm of day and night is created, to which people are accustomed. The cylinder revolves slowly around its longitudinal axis. The people therefore stand up with their heads in the direction of this imaginary axis. Each inhabitant would be able to see his or her 'antipode' to the left and right in the two other landscapes, were it not for the fact that the air and the clouds (which can develop if sufficient moisture is developed spontaneously) would in all probability obstruct the view.

The landscape can also be arranged as desired in this cylindrical world, like, for instance, densely populated areas or open agricultural

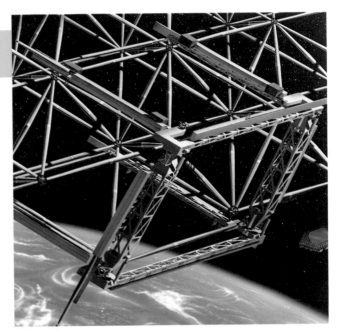

A solar power satellite under construction.

Top, left: A space mechanic during the assembly of a large platform for the reception of solar energy.

Below, left: Standing outside, one can see 'the ends of one's earth', to left and right.

Centre: The principle of a solar power satellite in space. The electricity derived from solar energy can be changed into short-wave radio beams. On Earth enormous antenna installations receive these beams, which are then transformed into electricity; for example, to supply a city.

land complete with trees, meadows, narrow paths, hills and small rivers.

What is the use of these space regions? The study groups in the United States already envisage a very clear and most important aim: the supply of cheap and clean energy to Earth and to their own miniature world. The inhabitants of the space colonies will mainly occupy themselves with the construction and maintenance of enormous solar energy centres (made of Moon materials) orbiting in wide paths around Earth. These centres beam electricity in the form of radio waves down to Earth, where it is converted into electricity again. The space inhabitants will also make products of great commercial value which cannot be produced on Earth, or only with great difficulty, due to the ever-present gravity.

Noteworthy for the space colony project is that it can be executed within currently available technology — if the money is released for this purpose. The development and construction of the very first space colony — the village for ten thousand inhabitants — would cost about one hundred million dollars. This is ten times the cost of America's first space station. At present, many of us can hardly imagine living in such a space colony forever. But the inhabitants of the space colony could well become very attached to a world without earthquakes and floods, thunderstorms and cloud-bursts. Perhaps they will come back to Earth for their holidays to 'do it the primitive way', if their stay is not for too long.

The biggest advantage of life in a space colony or a space city over life on the outer crust of a natural planet is that the environment of the artificial world can be controlled precisely. No unwanted freaks of nature, therefore. It may be that our descendants in the space regions will decide to have some snow at Christmas. Then this will be a democratic decision. This does not imply, however, that the Moon and the planets could not remain reasonably attractive places to people with the true pioneering spirit. In the beginning, there will be interest only in the basic materials, but later, when more facilities are available, permanent settlements will emerge. The situation can be compared with that of the early American pioneers, whose privations were often the spur to start small communities, which later grew to become major cities.

Cavemen

There is no doubt that the Moon will be the first to be permanently inhabited. It is nearby (compared to the planets) and it is a source of the materials required for the large space constructions. The Moon is an ideal place for ore production. But this will necessitate taking special steps to keep alive the people who live and work there. It would not be feasible for people living on the Moon to hop around (just like the Apollo astronauts) dressed in space-suits, particularly if there is work to be done. Undoubtedly the production of ore will be highly automated, but the first Moon colony will have to consist of at least 150 persons. They must be able to move freely within the artificial atmosphere of their control centre, to operate the automatic installations by remote control. A relatively simple solution is offered by an inflatable dome, similar to the domes over open air swimming pools in cold weather. A dome like this is kept inflated by its internal pressure, and an airlock will ensure that on entering and leaving there will be no loss of the precious atmosphere.

An additional, but not unimportant, aspect of structures of this nature on the Moon is the need for 'armour'. A simple plastic dome could be easily penetrated by small meteorites, something we on Earth are protected from by our one hundred kilometre thick atmosphere. We shall have to look for another solution for use on the Moon. We could cover the dome with Moon dust, but this would mean that the construction would become considerably heavier. The best way round the problem could be found in the natural cavities in the Moon, or in underground caves created artifically. Inside these caves, balloons could be inflated in which the domestic sites for the inhabitants are built. It seems that on the Moon, too, the first inhabitants will be cavemen.

Plenty of energy

There is no problem about energy on the Moon. Wherever you are (excepting some small areas around the Poles) the Sun shines undimmed for 14 days, to disappear for another 14 days. It is easy, therefore, to generate large quantities of electricity with the aid of solar cells. Part of the energy can be stored in batteries for use during the night. There is another method of generating energy, which may perhaps be preferred. In the long term, solar cells will be damaged by meteorites. So the other method is . . . the 'steam engine'. This would be very attractive on the Moon, with the difference that the classical water would be replaced by nitrogen. This gas would have to be supplied from Earth in liquid form but once it is on the Moon, it can last for an eternity. Solar heat is used to transform the cold liquid into gas, which drives a turbine, subsequently to condense into a liquid again. And that liquid is transformed into gas again, and so on and so on. This process is possible due to the absence of an atmosphere on the Moon: not only is it very hot in the Sun, but it is also very cold in the shade.

Not only to the 'gold panhandlers' of the coming era will the Moon be important, but scientists will be highly attracted too. For instance, for the construction and operation of a radio telescope at the back of the Moon, that side which we never see from Earth. Here, free from earthly

If man returns to the Moon it will not be just to plant a flag, but to stay there forever . . .

Extreme left: This is how the flights to the Moon started at the end of the sixties, beginning of the seventies. This was the only night launch of the Apollo programme - Apollo 17, for the time being the last journey of Americans to the 'Guardian of the Night'.

Inset: Lunar modules and lunar rover of the Apollo-16 flight.

Left: One of the Apollo-17 astronauts collecting moon-dust. Thanks to the research on this material we now know that practically all (building) materials are available on the moon.

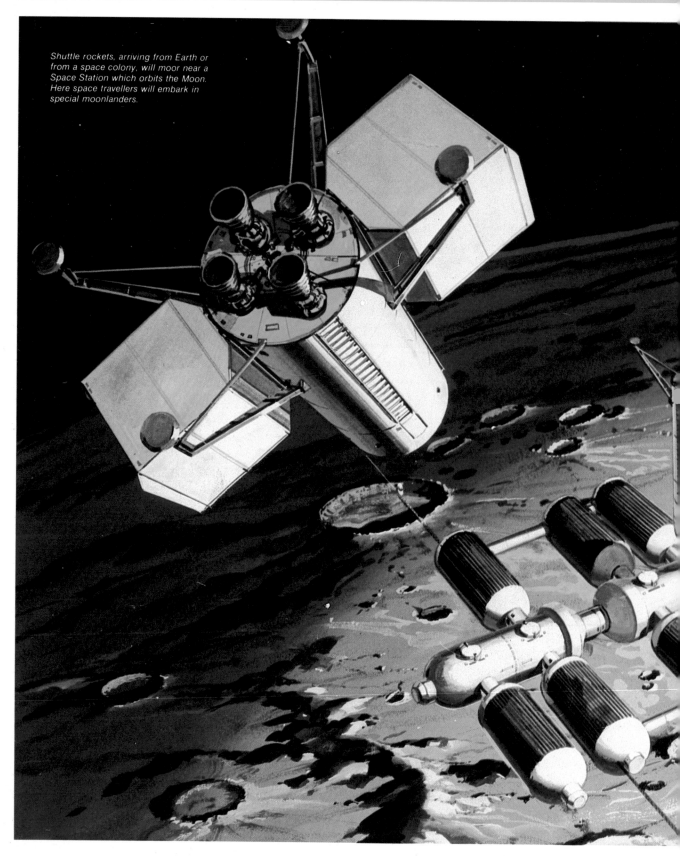

Shuttle rockets, arriving from Earth or from a space colony, will moor near a Space Station which orbits the Moon. Here space travellers will embark in special moonlanders.

Top, right: A temporary dwelling on the Moon, consisting of an inflatable dome and some small moonlanders.

Below, right: An underground Moon base designed for twelve people, which is adequately protected against meteorites and dangerous radiation.

interference, they can listen to signals from deep in the Universe. Perhaps signals from other intelligent beings. The low gravity also allows the construction of very large optical telescopes without imperfections in the reflector, which normally reduce the advantages of the large size. The biggest single reflector telescope on Earth has a diameter of 20 ft (6m). But on the Moon a reflector of 165 ft (50m) could easily be built. With this the Universe could be explored unto its farthest depths. And the planets of our solar system could be measured almost minutely, also due to the absence of an image-distorting atmosphere.

FUTURE: LIVING ON VENUS AND MARS

where temperature and pressure are quite acceptable, with a pressure of one kilogram per square centimetre and at about room temperature. Algae are plants and they will, according to Sagan, consume the carbon dioxide eagerly, because they love it. In doing so, they will produce oxygen (just like all plants), the gas so essential to us for life. By constantly improving the composition of the atmosphere the algae will conquer the lower regions until, ultimately, the whole atmosphere is almost free from carbon dioxide and full of oxygen.

Subsequently micro-organisms will have to be used to rid the atmosphere of Venus of an excess of sulphur dioxide, which otherwise would be responsible for a lot of very

Although the planet Mars will in all probability be the first heavenly body for Man to set foot on after the Moon, Venus can perhaps be made suitable for us for large-scale habitation earlier than Mars.

It is true that Mars is at present a much more pleasant (or rather less unpleasant) planet than Venus. But the low gravity on Mars (not quite 40 per cent of that on Earth), and because of this its extremely thin atmosphere, makes it difficult to turn this planet into a subsidiary of Earth.

On the other hand, Venus, in size and gravitational force, is almost the twin of Earth. But it has the 'misfortune' of being closer to the Sun than Earth is. This makes its surface hotter, and liquid water is absent. Thus the carbon dioxide gas originating from the volcanos is not bound in the rocks, like it is on Earth, but remains in the atmosphere. Carbon dioxide has a strong greenhouse effect: the Sun's heat can penetrate to the surface, but that surface cannot get rid of its radiating

heat via the atmosphere. Result: on Venus it has become hotter and hotter.

Space robots have allowed us to peek at the surface of Venus, which is permanently covered by a thick layer of cloud and so invisible to the eye: a boiling world, in places glowing dark-red beneath an infernal atmosphere.

The atmosphere on Venus is almost completely carbon dioxide — Nitrogen and oxygen are virtually non-existent. The pressure is nearly one hundred kilograms per square centimeter, which is about the same as one kilometre deep in Earth's oceans.

This would make habitation almost impossible. But, don't panic, aspiring Venus dwellers! The famous American astronomer Dr. Carl Sagan has developed an extremely clever plan to deliver Venus quite soon from this horrendous situation.

He wants to 'seed' the planet's atmosphere from above with a tough species of algae. High in the Venus atmosphere there are natural regions

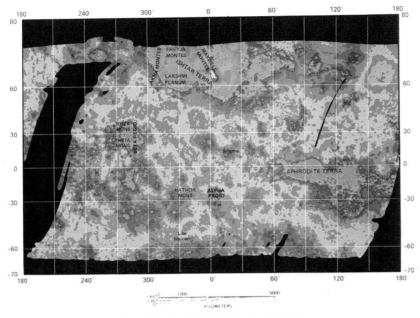

PRELIMINARY TOPOGRAPHIC MAP OF VENUS
CONTOUR INTERVAL ½ km ALL ELEV REFER TO A RADIUS OF 6045 kilometers

Opposite, top: What Venus looks like without its cloak of thick cloud. Perhaps this view is something for the future.

Below, centre: An impression of a large ravine on Venus, in the eastern part of the continent called Aphrodite Terra, 2250 km long and 280 km wide. The valley is 3 km deep.

Left: A map showing a large part of Venus. Differences in height are indicated by the various colours; the blue means lower, the yellow (red) means higher. The high continents of Venus are clearly shown, as well as the highest mountain 'Maxwell Montes'.

Below: A Russian Venuslander on the boiling hot surface of Venus. The temperature is 500 degrees Centigrade, the atmospheric pressure more than 90 atm. This impression was painted by the Russian artist Andrei Sokolov.

Left: Venus' largest continent 'Aphrodite Terra', which is about half as large as Africa. The highland is about 6,000 miles (9,700 km) long and 2,000 miles (3,200 km) wide. The area has numerous mountains and ridges which reach up to 2 miles (3 km) above the surrounding terrain.

Below, left: The Maxwell Montes on Venus is higher than Mount Everest. This mountain reaches to 6.7 miles (10.8 km) above the average surface height of Venus.

Below, right: The volcanic Maxwell Montes mountain range seen from above.

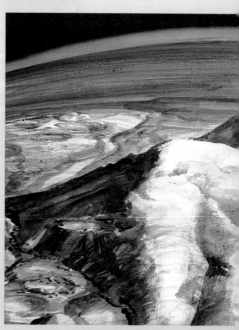

unpleasant acid rain. Once this is taken care of, Venus, which looks much like Earth in every respect, could quite easily be prepared for habitation — certainly in the areas around the Poles, where the Sun is always low. In the region of the Equator it would be far too hot, because Venus is much closer to the Sun than is Earth.

Loud laughter was the initial reaction to Sagan's plan. They thought it a crazy idea . . . until experiments in an imitation Venus atmosphere showed that the algae did their work even more quickly than Sagan could have hoped for. So it might happen that Venus, perhaps in the next era, but certainly in the subsequent era, could become a habitable planet: perhaps within a few years after the start of Sagan's project an acceptable environment could exist on the planet; at least in wide areas around both Poles.

Mars

As far as Mars is concerned, we will not get away with it so easily. The atmosphere around the 'Red planet' is

An automatic Marslander, having scooped up Martian soil, returns to Earth with its booty.

ten thousand times thinner than that around Venus. The atmospheric pressure here is only about 4 millimetres per square centimetre while on Earth it is 760 millimetres per square centimetre. The temperature on most of Mars is some tens of degrees below zero, but along the Equator there are temperatures well above zero. In contrast to Venus, the temperatures do not pose a big problem to us. The situation is more or less comparable with that near the South and North Poles of Earth.

But the thin atmosphere will inevitably necessitate the same far-reaching measures as those which we had to take on the Moon. Mars is somewhat heavier than the Moon but does not have sufficient mass to have a really dense atmosphere like Earth.

Earlier it was believed that Mars' atmosphere could be made denser by melting the polar ice-caps, so releasing gas. These polar-caps mainly consist of frozen carbon dioxide, like the carbonic acid 'snow' in fire extinguishers. The northern

polar-cap apparently also contains some water-ice. By scattering soot on the caps they would absorb more heat from the Sun, the carbon dioxide and the water-ice would melt, and water vapour would be released into the atmosphere. But it seems that we can forget this plan, because the polar-caps on Mars are so thin that there is not much chance of all this happening.

To actually change Mars would therefore appear impossible. But undoubtedly Mars will see the erection of small, and then larger (scientific) bases in the next century, some of which will probably be safely embedded underneath the pink deserts of this planet, which appeal so much to our imagination.

While a Viking parent-ship is taking photographs high above Mars, a lander is investigating the surface.

Left: A Russian capsule (Mars-3) was the first to make a soft landing on Mars. Impression by Russian artist Andrei Sokolov.

Below: This photograph, taken by the Viking-2 lander, shows a part of the lander itself as well as part of Mars.

The first colour photograph taken of the surface of Mars. The colouring is due to the presence of iron oxide. The picture was taken by the American Viking-1 in 1976. The hollow rock on the right is called 'The Dutch Clog'.

This is what a future landing on Mars could look like.

Below: Sunset on Mars, and the scoop used by the Viking to take samples of the soil.

In-between the orbits made by Mars and Jupiter, at least fifty thousand minor planets describe their orbits around the Sun. These minor planets, sometimes called asteroids or planetoids, are true miniature planets: crumbs, as it were; left-overs from the creation of the solar system. They could be the remains of a large planet which at one time travelled in an orbit between those of Jupiter and Mars, but which — in the nebular past — exploded. (According to one fantastic theory, this planet was populated by irresponsible chaps who were slightly negligent in the handling of nuclear power.) The vast majority of these minor planets are between a few hundreds of metres and few

kilometres in diameter. Many are smaller, a few are larger. The giant among the dwarfs is Ceres, a pock-marked potato with a diameter of 800 kilometres. If one could stand on this miniature world and kick a soccer ball it could easily disappear forever into the infinite, which shows what little gravity there is on such cosmic crumbs. That minor planets will ever become domiciles for people is, therefore, highly unlikely. But there is a fair chance that they could be used as sources of raw materials. This would be a very logical sequence to the exploitation of the Moon. Because, as we extend our activities in space, our demands for raw materials will increase accordingly.

Some thinking has been done concerning so-called 'mass-rockets'. A rocket like this could take a minor planet in tow into the vicinity of the Earth and the Moon, consuming a small portion of the miniature planet for fuel en route. This fuel could be processed during the trip. The minor planet would meanwhile have been wrapped in a large airtight bag, in which the miners can do their job, without being hampered by cumbersome space-suits. This method would provide us with tens of thousands of planetoids, which to us would form an almost endless source of material and energy. The Club of Rome evidently shouted too soon that there were limits to expansion,

because even energy problems in space are non-existent: the Sun keeps on shining.

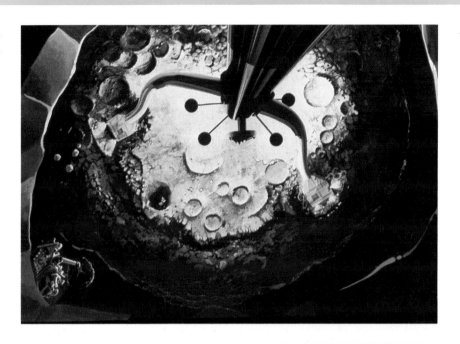

Top, right: 'Space booty'; an asteroid being wrapped up prior to its transport to Earth or Moon by rocket (foreground).

Below, right: A neatly wrapped asteroid passes the Moon en route to the ore-processing space factory.

Below: Jupiter is the largest of the planets in our solar system. It would hold a thousand earth-sized spheres. The large red spot is a type of permanent hurricane in this giant planet's atmosphere.

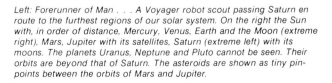

Left: Forerunner of Man . . . A Voyager robot scout passing Saturn en route to the furthest regions of our solar system. On the right the Sun with, in order of distance, Mercury, Venus, Earth and the Moon (extreme right), Mars, Jupiter with its satellites, Saturn (extreme left) with its moons. The planets Uranus, Neptune and Pluto cannot be seen. Their orbits are beyond that of Saturn. The asteroids are shown as tiny pin-points between the orbits of Mars and Jupiter.

Building material

The gigantic planets Jupiter, Saturn, Uranus and Neptune cannot be used for settlements, as we learned from our space-probes. They are enormous globes of primarily gas and liquid (hydrogen and helium with, perhaps, a small, solid core). There is no firm surface as on those planets nearer the Sun.

However, space prophets foresee a use even for these giants and their numerous moons: they could serve as building material for the construction of a whole series of replicas of Earth, which in the future could be circling the Sun like a swarm of industrious bees, warming themselves on the energy from this central star. At first sight this seems quite mad. And at second sight it may still do so. Perhaps we shall never get to the stage of realizing this kind of science fiction. But Jules Vernes was called a fantasist by his own generation (and by at least one more generation after that). However, it is useful and inspiring that some people look a long way past the ends of their noses. In

This photograph of the whirling, ice-cold Jupiter atmosphere looks like an abstract painting.

doing so they point out the opportunities to their less imaginative contemporaries and their descendants. A scientific dreamer (in the nicest sense of the word) like this is the American professor, Freeman Dyson. He suggested that in the far-distant future a cloud of artificially produced worlds could whirl around the Sun. In this way, solar energy could be put to better use, because the small planet Earth on its own only catches about one billionth of that energy. Such a cloud of worlds is called a Dyson-sphere. Dyson is of the opinion that every technological culture — and there may be billions of them in the Universe — will eventually leave its original planet, and in the end will become dissatisfied with the materials available around its own star. At that time the pioneers of such far-advanced cultures will travel to nearby stars which seem suitable for exploitation of the surrounding galaxies.

Unimaginable

The poor layman, of course, will have great difficulty in following these far-reaching (and perhaps far-travelling) futuristic visions.

Stumbling block number one is the distance between the stars, which staggers one's imagination. If we can think of the Sun shrunk to the size of an orange, which we subsequently put down in London, then Earth (as big as a pin-head) would describe a circular orbit of 25 ft (7.50 m). But the nearest star would still be as far away as Moscow!

The Sun's Planets

Name	Distance to Sun in AU (*)	Orbital period in years	Diameter in km	Rotational period in d, h, min	Average density	Known number of moons
Mercury	0.39	0.24	4,880	59d	5.4	–
Venus	0.72	0.62	12,105	243d	5.2	–
Earth	1.00	1.00	12,757	1d	5.5	1
Mars	1.52	1.88	6,753	24h 37 min	4.1	2
Jupiter	5.21	11.86	142,800	9h 51 min	1.3	15
Saturn	9.52	29.46	119,300	10h 41 min	0.7	18
Uranus	19.88	84.01	51,800	10h 48 min	1.25	11
Neptune	30.06	164.79	49,500	15h 48 min	1.67	2
Pluto	39.44	247.69	3,300(?)	6d 9h	1(?)	1

(*) = AU (Astronomical Unit) is the average distance Earth-Sun: 149,985,000 km

It is almost unimaginable. But maybe it will be possible for us to undertake journeys between the stars. A civilization which has reached the stage of being able to control a complete collection of stars could, according to Dyson, even explode stars in order to obtain basic materials. In that stage we would become independent from our Sun, which only has another five billion years ahead of her. The life of a star often ends in a gigantic explosion in which a former relatively weak star can become a hundred thousand times brighter. Every now and then a seemingly new star (previously too weak to be seen from Earth) flares up in the night skies. We call this a nova. According to Dyson it cannot be ruled out that this kind of catastrophe could sometimes be a living sign of a civilization which has achieved a truly cosmic level. If we humans get as far as that, we shall only know Earth from handed-down stories, Only then shall we have become true inhabitants of Space.

Next page: Saturn and its most important moons. This is an artist's impression of real photographs taken by the American Voyager-1. In the foreground the moon Dione; right, below Saturn, the moons Tethys and Mimas; to the left, Enceladus and extreme left, Rhea.

Below, left: The Voyager passing Saturn, taking photographs as it sails past.

Below, right: For the first time a volcanic eruption was observed away from Earth – on the orange moon Io of Jupiter. This boiling world appears to have such eruptions continually.

GLOSSARY OF TERMS

Apogee The highest point of an orbit around the Earth.

Azimuth The angle between a fixed point on the horizon (north or south) and the direction of travel.

Blackout The cessation of communications during the return into the Earth's atmosphere, due to the formation of an electrical charge in the ambient air around a spacecraft. Also: loss of consciousness during high accelerations.

Countdown The final phase before a flight commences.

ESA European Space Agency.

ET The External Tank of the space shuttle, containing liquid hydrogen and liquid oxygen.

EVA Extra-Vehicular Activity, sometimes called a space walk.

Fuel cell In which hydrogen and oxygen are combined to become water, during which electricity is produced.

G-forces Forces caused by acceleration (change in speed); expressed in 'g', 1g being the acceleration due to gravity on the surface of Earth.

Geostationary orbit An orbit at 22,000 miles (36,000 km) above the Earth's Equator. Here the orbit's duration is 24 hours, precisely equal to the Earth's rotation around its axis. A satellite located in this position will maintain this position with respect to a point on the Earth's surface at all times.

Hypersonic Speeds of more than Mach 5 — more than five times the speed of sound.

IUS Inertial Upper Stage; a rocket used to lift satellites to extremely high orbits.

LOS Loss of Signal; cessation of communications with the space vehicle.

Mach The speed of sound. At sea-level = 745 mph (1,200 km/h).

MET Mission Elapsed Time; the time elapsed since lift-off.

Micro-gravity Very low gravity caused by attraction of objects in orbit. The popular term is weightlessness, although true weightlessness can only be achieved by a body if it is immensely far away from any other bodies.

Mission Control The Control Centre for space flights.

MMU Manned Manoeuvring Unit or 'space scooter'.

NASA National Aeronautics and Space Administration, the American Space Agency.

OMS Orbital Manoeuvring System: two rocket engines which are part of the space shuttle, used to change the direction of travel.

OPF Orbiter Processing Facility, the building in which the spacecraft (orbiter) is checked after a flight, and is prepared for the next one.

PAM Payload Assist Module; a rocket engine used for taking a satellite to a higher orbit.

Perigee That part of an orbit which is nearest to Earth.

Pitch Upward and downward movement.

PLSS Portable Life Support System; a box carried on the back of an astronaut, necessary to stay alive in space.

RCS Reaction Control System; a number of small rockets used to control the attitude of a spacecraft.

Retro-rocket A rocket engine fired in the forward direction of a spacecraft, so reducing the spacecraft's forward speed.

RMS Remote Manipulator System; the manipulator arm used (among other purposes) for launching and retrieving satellites.

Roll Movement around the longitudinal axis.

STS Space Transportation System; the complete spacecraft transport system, consisting of the spacecraft (orbiter), booster rockets and the external tank.

Yawing Movement about a vehicle's axis, from side to side.